CW00717884

Honey Days

HONEY DAYS

by
OLIVER FIELD

illustrated by
ELIZABETH O'ROURKE

NORTHERN BEE BOOKS
Mytholmroyd : Hebden Bridge

ISBN 0-907908-55-1

Published by Northern Bee Books, Scout Bottom Farm, Mytholmroyd, Hebden Bridge, West Yorkshire.

Printed by G. Beard & Son Ltd., Brighton

CONTENTS

Preface		7
Chapter One	January	9
Chapter Two	February	14
Chapter Three	March	22
Chapter Four	April	27
Chapter Five	May	33
Chapter Six	June	40
Chapter Seven	July	48
Chapter Eight	August	57
Chapter Nine	September	65
Chapter Ten	October	72
Chapter Eleven	November	76
Chapter Twelve	December	79
Conclusion		85

ILLUSTRATIONS

January	Green Woodpecker sitting on Hive	10
February	Bees working Willow	16
March	Green Plover with Eggs	24
April	Chiffchaff on Cherry Blossom	29
May	Cuckoo on Horse Chestnut Bough	34
	Trout taking Mayfly	37
June	Spotted Orchids	40
	Thames Barbel	45
July	Frog and Water-Lily	51
August	Sainfoin Flower	59
September	Buzzard over Dartmoor	69
October	Barn Own and Corn Rick	73
December	Rabbits in Snow	81
	Rooks in Flight, right at the end	83
Conclusion	Bumblebee on Dandelion	86

PREFACE

Oliver Field invites you to join him as he tends his apiaries around southern England. You will pick up a lot about beekeeping but you will also learn a great deal about our countryside, changes in farming methods and the impact which this has had on wildlife. From the moment he could walk, Oliver has been an unashamed enthusiast for the simple pleasures of the countryside. He will share with you the quiet moments watching swifts catch mayflies as they rise from the river, regular visits to a favourite patch of orchids, and he will recall the thrill of landing an eleven pound Barbel.

I share Oliver's enthusiasm but as a scientist and a professional conservationist, I am expected to embrace only unemotional objectivity. My colleagues and I assess the resources of nature, evaluate it in the most objective way possible, and ensure that we have evidence to back up the actions we take. The need for scientific respectability makes it very difficult to provide convincing data on habitat decline over a period of yeas or decades. It is not that there is any doubt about the enormity of the change, but people like to be able to quote verifiable statistics. The main problem is that before the recent agricultural revolution, our heaths, downs, meadows and coppicies were taken for granted. Habitats and species were not surveyed and documented to provide baseline data. When you have read Oliver's book you will probably say 'Who needs data?' Do we need any better evidence than the reminiscences of the observant countryman who has been taking an intelligent and sensitive interest in his surroundings since the 1940's?

This book recalls so many changes that the reader can not be left in any doubt that our countryside has been depleted of a high proportion of its character and diversity. This is not apparent to everyone, as the countryside looks broadly unchanged through the windows of a speeding car or train, but to those who have troubled to stop and ponder, there has been a depressing erosion of nature's intimate beauty.

This century has seen a startling reduction in the proportion of people whose lives are essentially rural. As society becomes more sophisticated we move further away from the earthy need to produce food. True, a great many people have moved into villages to get away from the bustle of the city, but they live in a very different world from the true countryman. There is a danger that we will forget thet we share the land with a great many other creatures and a host of wild plants. We need people like Oliver to demonstrate the pleasure which can be derived from nature, not just in the special places but on our very doorsteps. Nature should not just be marvelled at on television, or only appreciated in rare visits to nature reserves. No one has to travel far to find a corner where nature can be admired or studied. But how many people bother? We must open our eyes, and find time to indulge in a little natural curiosity. The current over-production of food, combined with concern for 'green' issues, gives us the opportunity to retrieve much of the damage of recent decades. This will only be realised with a high degree of political will, and this only comes if enough people really care.

A beekeeper has to be extremely aware of the seasons, the annual cycle of farming operations, and the way that nature responds. A professional like Oliver is almost uniquely placed to comment on the health of the countryside, for he travels widely, he visits many habitats and is in touch with many of the people who dictate what happens on the land. The pages of this book contain far more than readable reminiscences. They provide a chronicle of events and changes which to most people have passed unnoticed but which we need to be aware of if future generations are not to be denied some of the basic ingredients of the quality of life.

Dick Hornby
Regional Officer,
South England Nature Conservancy Council

Chapter 1

January

The great joy of being a honey farmer is that one lives one's life with the changing seasons; every day is different, each day carries its own delight, as we wander from place to place or county to county.

January to me is indeed the depth of winter, the start of a new season. A few years ago I went to Scotland. I went there to look at a honey farming project that the Highlands and Islands Development Board envisaged, on the Black Isle. On my journey north I stopped at Newburgh in Fife and spent the night with some friends by the Tay Estuary.

At a very early hour, long before dawn, we took a boat and went across to the dog bank, a huge mud flat on the north side of the estuary. I have to admit I took a gun and settled down in a clump of short reeds on the edge of a deep, muddy gully, to wait for the greylag and pink-footed geese to flight in at dawn. My friends took the boat away to another clump of reeds on another patch of mud.

As I sat there on a far from dry sack draped over a short plank of wood, I watched the turning of the tide. It was very cold, a white frost clung to the tops of the reeds, the water lapped gently a few yards away. this was the very depth of winter, it was the first week of the year. Platforms of ice were drifting down the river with the outgoing tide. High above I heard the mallard returning from their night time forays inland. Then, as I watched, I noticed that the ice flows had stopped their drift to the sea. They circled slowly in the water, this was the moment of low tide. Within minutes the water would begin to creep back into the gullies, the ice would change direction and start its course up the river again. Imperceptibly the change had taken place, and at the same time a gathering lemon flush was creeping into the sky. Dawn would soon be with us.

The change in the tide, the change of the seasons, both so intricately intertwined in so many ways. The pull of high summer was far, far away, but in that week, maybe at that moment, the direction was changing. Like the incoming tide, spring would soon come swirling into view, swelling the buds, thrusting the grass forth, and

Green woodpecker sitting on a hive

waking the bees from their long sleep. I sat and watched the mud disappear as the gully filled. I saw the ice flows return, now covered with mallard standing and preening themselves, or asleep. The sky was soon much brighter, the call of the geese rang out up and down the estuary, the armies were on the move. It was a perfect still morning with a bright blue sky, but bitterly cold. The geese passed high above my head, far out of gun shot, their grey and white shapes etched against the sky. I was happy to be there, and not sorry to return without firing a shot.

For me the great love is to be in the country, to be free, to live within the natural world as far as I can, and honey farming gives me that freedom. I can watch and see, I can stand and stare, and live those moments that matter without interruption or disturbance.

I ask those who would care to come with me to spend a season out with the bees, to learn the pulls of the countryside and see for yourself what really matters. We will wander day by day to new woods, rivers and streams; to fields full of flowers, to moors that are a mass of heather. If you come with me you will forget the office and listen to the first chiff-chaff on the twenty-eighth of March. You will be late home when we move the bees to the apple orchards. You will hear the nightingale before the last hive is off its stand, when we take them back to their summer sites.

January is a bleak time of the year. The leaves are all off the trees, the wind can find you wherever you are. At dawn and dusk a few hardy mistle thrushes will sing and a fox may bark in your apiary on a moonlit, frosty night. I have stood amongst the hives on just such a night. I have walked through the woods and looked at those dark shapes sitting on their stands and thought of the life within them. That ball of living creatures, tightly packed together; the ones on the fringe get cold and burrow back into the warm centre, while the warm ones get pushed out, and so on and so on. Bees need very little looking after at this time of the year. Green woodpeckers may represent a problem if it gets very cold and they can't find the ants that they generally eat.

We may get a day in late January when the weather relents, when there is a gently south wind, and the sun shines in the middle of the day. The heat is barely enough, but for the first time in several weeks the bees break cluster and fly from the hive. The first snowdrop unfolds her petals and hangs daintily by the hedge. For a moment a bee will hover among the heads, then the sun will go behind the clouds and we will be back to winter.

I think that it would be true to say that the growing of oil seed rape has revolutionised honey farming in the UK. These days we rely on several tons of this crop every year. When I first became a honey farmer, it was the spring sown crop that made up the lion's share of our summer honey, but after the dry summers of 1975-76 there was a change and we began to see more and more winter sown, spring flowering rape. Today, with new varieties, this flowering period can extend from early May right through to July, and in 1989 we took six tons of honey in June.

With this upsurge in rape growing, the wood pigeon has also flourished and its population has gone steadily up and up over the last twenty years. One of the problems for the wood pigeon is to find enough food during the winter months. He will strip the clover fields in October and November and feed on the beech nuts and acorns in the fringe of the wood. When January comes the cupboard is bare, the frosts and snow carpet the fields, and the wood pigeons become very hungry. It is then that the oilseed rape will keep them alive as it peers through the snow. Unfortunately this is not good for the farmer or the honey producer; leaves are shredded and the fields stripped bare of any green leaf.

It is now in January that the bee farmer will find himself sitting in a straw bale hide with his flask of hot coffee, keeping the grey hordes away from the fields. It is at this point that my reader will say "this man is always shooting at things", but please bear with me. I love pigeons on hot summer days as they call from the woods and flap slowly across the rising barley to drink from the water trough in the corner of the field. On the other hand, I have seen the damage that pigeons can do to a field of winter rape.

You can come and sit with me if you like. You can watch them coming towards us out of the firs over the winter wheat, along the top of the hedge and across the rape. The farmer has asked if we would like to shoot pigeons. He is delighted we are there. "Bee farmers are idle people in the winter", he will say, "so why shouldn't they protect my rape fields, they might just as well sit out there in the frost and cold as sit at home by the fire, at least they are doing some good out there and they may even get a crop of honey in the early summer".

Pigeons are not easy to shoot. They have very, very good eyesight, and as they cross the field they look all about. There may be twenty pairs of beady green eyes watching for a movement. You lift your head too soon and they are gone, turning on their great broad tails, twisting away with their wings, and in that second your shot goes wide. Wait too long, or shoot too soon, and you will always miss far more than you take home in the bag, but the farmer will be happy. He will hear the shooting and he will think well of you and of your bees the next summer. There is always so much to see as you sit and wait. It may be the movement of the carrion crows that will rarely pass over you, or the high flight of mallard hurrying up the Thames Valley. Did you see that grey-brown bird like the pigeon but with his head hunched back on his shoulders, skimming low along the hedge? It was a cock sparrow hawk. Did you see the way he dashed through that gap in the hedge? No doubt he saw a blackbird or a thrush on the ground beyond.

When I was a boy there were many more sparrow hawks than there are today, but from the mid-forties to the mid-sixties they became very scarce. They were poisoned by grain dressings which had been eaten by the birds that they preyed on. Luckily this was noticed in time, and now the sparrow hawk has returned. Once, as a small boy of seven, I was walking the meadows around Warfield when I saw a huge flock of fieldfares descending on a high, thick hedge that was loaded with hawthorn berries. Their chattering echoed all around me as they tucked into the berries. Suddenly a bird streaked over my head and plunged into the bush ahead of me. As he turned to fly off I could see a fieldfare clutched in his talons. At that moment he saw me standing not eight feed away. He dropped the migrant thrush and shot back over the hedge. I looked down and there at my feet was the victim; blood ran from a hole below its eye and stained the frosty grass. It gasped once as I picked it up and then it was dead. I turned it over in my hand, it was a beautiful bird with handsome markings, but so had been the sparrow hawk. It was probably my first lesson in life and death.

While I was talking a dozen pigeons passed to our right, but look – here come another six straight towards us – they have set their wings and are floating over the field, If you want a pigeon pie, now is the moment. If not, they will turn and fly slowly on looking for other pigeons to join and feed in safety.

By mid-afternoon the cold is returning, the warmth of the January midday sun is indeed short. The pigeons are now flying back to the fir trees or the thicket of the ivy in the oaks. The bees may have moved for an hour or two but now, like the pigeons, they will settle for the night. We leave the rape field with the mud caked on our boots.

Chapter 2

February

February is a fickle month. I have known lovely days in the middle weeks when one might have believed that spring had really come, but then again I have known days when all is white and the frost chills the very bones of our hives. You can never trust February. The mistle thrush that sang so lustily on those mild days in January now sits sullenly in the ivy, fluffed into a round ball to keep out the biting cold, and digests that last crop full of berries from the hawthorn. Such is February; no doubt in my mind – it is the coldest month of the year. It is now that our hives are ravaged by the weather. There is nothing you can do to help. The bitter cold freezes the bees onto their combs, they cannot move to fresh food and even if you feed them all the sugar syrup that they might need they still may starve. The problem is one of mobility, for if it is too cold they just stay where they are, they use up the food that they have beneath them and when it is gone they will be quite unable to reach any more. If the cold continues they will starve, within inches of their larder. If bees are strong, with a large cluster, then they will make it through the winter cold, but if they are weak stocks with small clusters, and late mated queens, then they will certainly die when the winter is very cold.

I have seen the first flowers of the wild plum break in the last week of January and I have seen the trees in full flower in a mild February, but I have also seen them still stark and bare at the end of April. You can never be sure how the spring will develop, each spring is different. In 1973 we hardly had a winter, it was mild right the way through, there were daffodils in flower at the end of January and my old wild plum showed its first flower on the sixteenth of that month. All was well until the end of March and then we got four inches of snow, followed by the coldest week of the year, and by then it was the first week of April. In 1986 it got very cold in the middle of January, we had the coldest night ever in mid-February and it remained raw until the end of April. As each season develops, nature develops with it and a balance is formed as we go along. Let us look again at our wild plum, the first fruit flower of the spring. If it blossoms in February it will need every flower it can produce, for the frosts will take the majority, but on the other hand, if it does not flower until the middle of April, then it will produce a heavy crop, so what matter if the bullfinches have had their fill of buds in February and March.

February is a month for all seasons. The weather is totally unpredictable, as it can swing from a glorious spring day back into the depths of winter in 48 hours. You can never be quite sure which way the next day's temperature will go, it may soar up the thermometer or dive down the scale out of sight. For a bee farmer this can be rather perplexing – one day the stock is on the wing looking for the first crocus, the next they are back in tight cluster huddled together to keep warm, such is February.

Not only can the temperature dive as low as you have ever known it, as it did some four seasons ago, but the wind can raise its strength to a hurricane and the heavens can open to deposit more inches of rain than you would care to imagine. No wonder we, as a nation, are obsessed with the subject.

"What are the implications for a bee farmer?" you may ask, and I will, of course, tell you for it is something that you really ought to know. Firstly, make sure that none of your apiaries are close to a stream or situated in low lying land. I have seen the most innocent little stream become a monster in three days and come rolling by, sweeping all before it in its haste to reach the Thames. Such little streams can flood fields, woods, roads, and if you are not careful they may carry your hives away on a giant orange wave. I have only once been caught, when I was quite new to honey farming and had not thought out my strategy as I should. The little stream, the Bull Brook, had been part of my life for 20 years and I had rarely seen it over its banks. I had fished it for minnows as a boy and I trusted it. I had no reason to doubt its honesty, so I placed my hives under some willows on a bend of the stream not 50 yards from the river bank. February came and so did the rains. I was not on the move in time and when I got to my site the road to the bees was already under two feet of water. I put on my waders and struggled across two fields to try to reach them from behind, but I could not, the water was too deep and there was a ditch between me and my hives. I stood and looked through the bare willows to the hives on their stands. They were sitting in water half way up their brood chambers with a pair of water hens clucking about between them. I had to wait a full week to reach them, and fully expected to find them all dead, but not a bit of it; the cluster seemed to have made sure that they kept their feed dry. As the water rose, so had

Bees working Willow

the bees, and I suspect that much of the cluster had come up through the feeder hole and onto the crown board under the roof. Believe it or not, I did not lose a single stock and by April they had cleaned their combs up and were happily flying about the willows.

February gales – now there's another thing to watch out for, especially nowadays when gales may turn into hurricanes without any warning to anyone. It should be remembered that the steadiest giant oak that you climbed as a boy can be torn from the ground and left a shattered heap on the field. These great trees weigh several tons and if one of them should come crashing across your collection of hives it will not do them any good at all. I have seen what can happen, and it is not a pretty sight. We had some bees under four huge beech trees on the edge of a wood, and in February a gale swept across the county. A huge branch from high up in the beeches became detached and fell right across the hives, scattering them in all directions. I got there just in time and put them all back on their floors, only one had been damaged. I cleared up the mess and repaired the stands. I had not been warned, and I should have, for only six months later, in August when the leaves were on the trees, there came another mighty gale and the force of it caused the trunk of the largest beech I have ever seen to break in half. It thundered through the apiary, throwing hives and supers in all directions. Three hives were smashed like matchboxes. I got there a day later to make sure all was well. It was a lovely warm sunny day, the bees were well on the wing. All I could see was huge branches of green beech, with thousands upon thousands of angry bees robbing supers and brood chambers alike. I had no saw with me, but I did my best, it was one of the fiercest bee battles that I have ever encountered. I was just not bee proof. I battled on amongst the branches, hunting for hives, trying to decide which super belonged to which hive, and in some cases picking up frames covered with bees all fighting each other and me.

So remember when you set up your little out apiary – trees are not eternal; they do come down. Watch out for the prevailing wind and make sure that it does not blow straight through that ivy covered oak that hangs over your bees, it could spell disaster.

Recently we had the gale to end all gales and I saw thousands of trees swept aside. I did not go very far while the weather was at its worst. I watched the fruit trees in my plum orchard rock in the soft soil and topple. I went for a walk, battling against the force of the wind. I watched a flock of wigeon speed across the sky and dip down onto a local gravel pit. No bird could fly against such a force. I wondered what my friends, the rooks, were doing. They could hardly be sitting in the trees, and to be on the wing would mean a trip across the land until they were far from home. I found them all sitting out in the middle of a grass field, and when I say sitting I mean sitting. All were facing the gale, beaks set forward, wings tightly tucked into their sides, bodies resting on the grass. Not a movement. They sat motionless in a tight group waiting for the storm to blow itself out. I walked on down to the gravel pits and watched the waves rolling down the surface to break at the far end. All the wild fowl were also well sheltered, tucked in on the side of an island shielded from the blast.

February cannot be trusted, I say it again; why, only three seasons back, those same gravel pits were frozen solid. A man could walk across them and not be in any danger. For weeks the cold persisted. That year the wild fowl were far away on the coast.

As for work, February is another quiet month for the honey farmer. It's a time for reflection and planning and little else. All you can do is worry and hope the bees will be all right. Besides the weather, there is another problem that is ever with us in the cold of February, and that is the green woodpecker. Now I love green woodpeckers, never let it be said that I would raise my hand against them; they are a beautiful bird that adds grace and colour to the garden in the drab days of winter. To see a green woodpecker searching for ants on the lawn while a watery sun battles its way through the mist is a pleasure indeed. On the other hand, to see the damage that green woodpeckers have done in my apiary in the cold days of February will turn the Jekyll in my nature to Hyde in a very few moments. Why they have to knock a hole in every single hive, I do not know. They are welcome to a few bees from one hive to keep them alive but when they do wanton and untold damage to all twelve it puts up my blood pressure on the spot. Like rogue tigers who eat men for pleasure, there are rogue woodpeckers who peck holes in bee hives just to keep warm on frosty days. The colder it is, the harder they peck!

I remember once telling a trout farmer what a disgrace it was to shoot herons and what a lovely bird a heron was, and how lost the countryside would be without them. It was only when he pointed out how lucky I was that herons did not peck holes in hives and drew my attention the larger beak size that I credited him with a point of view. "Remember", I have to say to myself as I stand knee deep in wood chippings, "there are times when we must all hold our hand and not simply empty a gun".

Having got that off my chest, it reminds me that there are one or two apiaries that I must go and see. I must go and check that the nets round the hives are firmly in place and that the mouse guards have not moved. I will take a gallon of creosote with me, for nothing will make a woodpecker's eyes water faster than a little wood preservative round the holes he is trying to bore. Perhaps you might like to come along, I know that it's rather cold and there has been a hard frost, but we will not hang about.

The first apiary we must visit is the one near Shiplake lock. I know that there is a woodpecker there, he has had a go at the hives these last four seasons. I think that it is about time he died of old age. I told the local gamekeeper as much when I gave him a pot of honey in the summer. Luckily gamekeepers today are more enlightened than their fathers were, and he told me how often he watched the woodpeckers flashing through the woods in the summer and how lost he would be without them.

My Shiplake apiary is on the edge of a small beech wood to the north with a thick, low hedge to the south – an ideal spot with all round shelter. Now in February all is still, no bees fly, while a thousand thousand beech leaves lie scattered around the stands, like so many cornflake packets emptied about the site. The frost is white on the curled edges of the leaves, creating the illusion even further, as if someone had sprinkled sugar on the flakes.

The bees look fine as far as we can see, and our green friend has still been able to find enough ants to keep him busy. The nets have not even moved, though the gale last week might have brought down a branch or two and bounced them off the hive roofs.

It will be a grand afternoon to walk up the river towards Sonning. The walk will keep us warm and it is a lovely stretch at any time of the year. I know it well from boyhood. Across the river are the fields of Burough Marsh where my great grandfather lived just over a hundred years ago. He painted many pictures of this area and little has changed in all that time. There is a large area of low-lying land in the middle of the fields, which is always flooded in winter. You could stand there at dusk tonight and you would hear the whistle of wigeon and teal, and you would see dusky shapes weaving in and out of the shadows, just as they did when my grandfather was a boy.

On the other hand, in many ways the river has changed beyond belief; it is no longer clean and clear as it used to be and the population of fishes has changed. When I was young the bleak was the commonest fish in the river, there were millions of them, and the pike, the perch, the chubb and the trout all lived on them. Now the bleak has almost disappeared. They lived on the insects that fell on the surface but today, with all the thousands of launches that pound up and down the river, the water never settles. It is always as thick as pea soup, the mud is for ever stirred up and the surface is coated with oil. The little bleak cannot live in these conditions, so he has gone. Another problem is the question of spawn, for when the fish have spawned in the spring then the boats come and stir up the eggs and wash them away. They cut off the weeds and again the young of the fish are lost.

Many years ago my father and I would walk up this stretch of the river and then fish it down from Sonning for pike. On a good day we would catch six to a dozen fish in the day, I think that you would be very lucky to do that today. I well remember one afternoon in February when I was a teenager; we had walked down the river to the islands above Shiplake hole and I was fishing a small roach along the sheathing that supported the bank and tow path. As I reached the river bend my large red pike float hung for a moment off a small clump of reeds. At that moment the water erupted all around it and a huge fish swirled on the surface. A moment later the float was gone, the rod was bent into a half circle and it felt as if I had hooked a huge sack of potatoes which was steadily crossing the river. For some time the tug of war continued, while my father made haste down the tow path with a landing net. Eventually some three feet of very cross pike appeared on the surface, shaking his head and showing some rather nasty teeth. On the bank he weighed fifteen pounds, and for a teenager it was the fish of my life. I had to take it home and eat it. My father cut it into steaks and cooked it in cider with a few herbs sprinkled about for good measure. All I can say is that for anyone who likes to chew soggy cotton wool, with small pins in it, try and eat a pike. Today I am happy to watch them in the river and not to try and pull them out.

I trust I have not bored you with my memories, but passing that spot on the tow path brought back thoughts going back more years than I care to remember. If you listen I think I can hear the Canada geese on the move. It is getting late and it is time we walked back to the apiary and the vehicle. The Canada goose is now one of the commonest water fowls on the Thames, but it was not always so. During the last war they started to appear on the river and in the early fifties I found my first nest on one of the islands that we have just passed. Over the years they have spread and spread and now there are many thousands of them up and down the river. I think that one of their great strengths is that they are just about as eatable as the Thames pike and so they are left alone. Here they come, over the pollard willows and straight towards us, spread across the sky in a huge line, all calling to each other not twenty yards above our heads; they really are a fine sight and they are off to spend the night on the gravel pits, where they will feel safe. On moonlit nights they will leave their roosts and come back to the fields to feed. I have been here on the full moon and stood out there on the marsh and waited for them to come back. It was very exciting to be amongst them when they did not know I was there.

Well, here we are back at the bees. The sun has long gone and the frost has started to settle on our windscreen, and tomorrow we have got to put all those roofs together and do a day's work in the bee shed – it's time we were away.

Chapter 3

March

By March the tide of spring will be pushing its way back into the countryside, there will be a steady flow of warmth and light returning inch by inch and day by day. At dawn and dusk the blackbirds and song thrushes will herald the day and sing long after the sun has set. At last we can start to think about doing a little work, we can spend three days going right round the bee farm to see what our winter losses are and bring the dead hives home. If it is mild we can take off the nets and mouse guards to give the bees a chance to fly freely.

March is a month of excitement and interest, tinged with the disappointment of our losses. Few apiaries will have come through free of any loss, most will have at least one dead stock, and there will be the usual sprinkling of weak hives.

You might find it possible to boost the weak hives with combs from the stronger ones and so save them, but often, if the strain is poor, it is a problem of evolution, and to maintain these hives may not be in the long term interest of the bee farm. Too often today man can be found interfering with evolution; it is so easy to say "I don't like these rules of evolution, this world is a cruel and unkind environment, we must change things for the better". What people fail to realise is that these rules of evolution have been around for a very long time. In fact, some five thousand million years, and if you start to change them what are you going to put in their place, and will it work? It is no good saying that the survival of the fittest will no longer operate, but still expect to get the best.

For too long man has expected the natural world to serve him, to meet all his needs, to feed him, to keep him warm, to allow him unlimited liberty, without ever looking at the price tag. Today people are turning over the label and beginning to look at the cost, and they are appalled at what it is all going to need to put things straight. "You cannot buck the system for ever", as they say, "and expect it all to come out right in the end". As a bee farmer I see the damage of the last thirty years every time I go out into the country to visit my bees, and more and more I wonder what tomorrow will bring.

One of the first jobs to do in March is to visit the heather colonies. These hives are left on their stands on the heaths of Dorset all through the winter. There are a number of advantages in doing this, one is that they can work the heather right up to the middle of October or even later, so giving them plenty of late pollen. Another is that the winters in Dorset are usually not quite so cold as they are in the Thames Valley area. A third reason is the large quantities of willow pollen that can be gathered early, coupled with the ever present pollen from the gorse. Gorse is one of the few plants that seems to be able to have flowers on it in every month of the year. Gorse pollen is orange and well loved by the bees, but as for nectar, I do not think that it ever produces a drop.

If you come with us for a day going round the Dorset bees in mid March, I am sure that it will be a pleasure that will stay with you for a long time. If the weather is fine, and the sun is with us, there is much to do, and the surroundings are quite beautiful. It is a time of change, the Brent geese on Poole Harbour will be thinking of being away to the north, so will the hen harriers, the males in their smoky grey plumage will wander the length of the country to the Islands of Orkney. Gulls on the coast will be gathering along the cliffs back to their nesting quarters.

Willows start to come into flower in late February, in most seasons, and they will be a mass of golden male flowers by the middle of March. We have a site for the bees at Corfe, a stream runs down a little valley to Poole Harbour and all down the valley the banks are thick with willows. They used to be grown by the fishermen for the making of lobster pots. In years gone by the withy beds were kept cut and the long wands of willow that grew were seasoned and used in the winter to weave pots. Today lobster pots are no longer made this way; they are made by covering a frame with net so the willows have been left to grow. What a sight they are in the spring, male and female trees mixed together, the golden male flowers mingling with branches of silver female buds. The bees love this feast, the scent of the pollen drifts through the apiary as they hurry away to gather, and come home covered with pollen. The male flowers give them the bread of life, while the female flowers give them the nectar. I have known a huge nectar flow at this time of the year, and once or twice there has been a flow in late March that could rival any other of the season. The only problem is that the bees are still very weak after the long winter, and although they may have built up well on over wintered heather pollen,

Green Plover with eggs

they are still few in numbers. Even so, there have been two years in the last ten when we have actually taken full supers of ripe honey from these bees before the end of April, and very good honey it was too. This honey, when subjected to a pollen check, showed a 96% willow reading, with a touch of prunus, no doubt the little wild sloe. The sad thing about willow flows is the fact that, after they are over, there is often very little else for the bees until the first flowers of the bell heather. This leaves a long gap of at least two months or a little more, and we usually have to move the bees away to pollination and take them back at a later date.

In March the harsh note of the carrion crow echoes across the meadows, mingled with the gentler call of the pee-wit.

When I was a boy, the green plover was a rare bird in the countryside, At the turn of the century it was shot almost to extinction by professional fowlers, who sold plovers for maybe a shilling a head. Then it was protected, I think in fact that it was the first bird ever to be protected, although its eggs were still taken and eaten. As I grew bigger, so did the population of plovers, until now there are several millions spread all over the British Isles.

As a boy I would search the water meadows of the Thames Valley for plovers' nests, continuously persecuted from above by the parent birds, who would rise and fall on the wing, tumbling about the sky and continually calling, as pee-wits do.

Nowadays there is a certain time in middle to late March when the flocks of plovers will break up and spread to their nesting areas. In February their armies can be seen covering the fields, maybe a thousand birds together. The flock will lift and swing about the ploughed land, changing from white to black and then back to white again. Then suddenly, when the weather is just right, they are away to the water meadows, to the fields of tufted grass with their wet gullies full of marsh marigolds, or king cups as they are often known. The birds split into pairs and cover the territory with slow, deliberate flight, their wing beats clearly audible and their plaintive call rippling away across the river.

I can stand for hours and watch them. It is part of spring when the plovers have paired, and by the 26th March the first egg will be laid. Finding the nest is another matter, it is well hidden right out in the open, and though that may sound like a contradiction, it is not. My son and I have spent many many hours quartering the ground, and it is easy to pass the nest by several times before you see it. Once seen it appears so obvious that you can't imagine how you missed it in the first place. As for the eggs, they are very excellent to eat, but nowadays they should be left alone. I have to admit that I have eaten quite a few, the white is transparent and the yoke very orange. The eggs are beautiful to look at as they lie in the little scrape in the grass or plough, all pointing to the centre. I have watched the carrion crow keeping a note of the plovers, and I am sure that many first clutches go his way.

For me, the calling plover's note in my ears, as I work the bees in some of our low-lying apiaries in the valley, has a sense of permanence that returns every spring, to bring back memories long since forgotten. Rooks have the same effect, and apiaries sited near a rookery are full of bustle with the birds calling to each other as they hurry back and forth.

In late March we try to get right round all our hives once more to check them and make sure that each stock has a healthy and laying queen. It is at this time of year that you will find queens have died or queens have turned drone layer; in other words she is only laying male eggs. This will, of course, mean that if she is left then the hive will die out as it will soon become full of drones with no workers. All such queens must be killed, and the stock united with another stock which has a good queen. There are always the little lots of bees which have struggled through the winter and are queen right, but have very few bees and very little brood. If such stocks are united with the bees from the drone layer, then the two will make one good unit, whereas if they were left to their own devices then both would very likely be lost. It is on this circuit of the farm that we try to clip the wings of every good queen that we see. There will be queens that had their wings clipped the previous year, and the year before, so it will just be the young queens born the last summer that will have to be caught. The clipping of their wings has the main advantage that you will then know how old the queen bee is in every hive you own. This is very important, for if in the next summer you see a queen in a hive running around with a full set of wings, then you know that something must have happened. The old queen will have died or swarmed. Another advantage will be that an old queen cannot fly off with a prime swarm if she has no wings, so prime swarms will not be lost, though the queens probably will.

Chapter 4

April

In the orchards the first flowers of the season will be the plums. I have known plums come into flower as early as the 20th March, and in 1973, when we hardly had a winter, and again in 1989, we were moving the first bees into the orchards by this date. On the other hand, in winters as hard as 1985 and 1986, there were no plum flowers open before the end of April, such are the ways of our springs. If the winter is mild we get a long drawn out pollination period, which stretches from mid March to the third week of May. With cold harsh springs, everything is concertina'd together, and the cherries flow from the plums and into the pears, and the apples are out before you are aware that the plums have finished. Such seasons give the bees no time to do their work, and after cold winters their numbers are few, but nature compensates and will give us fewer late frosts, and so the crop may be good. In the early seasons the good early set of fruit may well get wiped out by an unkind night about 12th May.

Fruit blossom time is a glorious time to be out with the bees. Somehow for me nature is at her best. In our part of the world, the wild cherries give a white splash to the woodlands, and trees that you hardly knew were there are suddenly the focus of attention, for us as well as our bees. The wild cherry is a fine nectar plant in a warm, easy spring, and for those bees that do not go to pollination, it is the first real mass of flowers that they get to work. Often we will stand on the edge of a still, bare wood listening to the first chiff chaff, and watch our bees returning laden from the feast across the meadows where the cherries stand opposite us. We go through the hives trying to find the queen, and there on the combs are the bees with cherry pollen on their legs dancing a honey dance, to draw others over the field to their success.

At this time of year, nature surges forward and, like our incoming tide, the trees seem to flower and turn green overnight. The bare woods of late March will be a hundred shades of green by the end of April, the migrants will come pouring back to our fields and woods.

The swallows will chatter in the cow sheds, the martins will flit around the eaves. In the hedgerow a grasshopper warbler will reel away his high pitched song long after the sun has dipped. For us there is not enough time to do all the work, we hurry from apiary to apiary, putting on honey supers, getting stocks ready to move, either to the orchards or to those fields of rape where we shot the pigeons in January. As I have said earlier, rape has transformed the life of the honey farmer; nowadays we can look for a first bite of the apple in May and June, and another one in July and August. When I was first a honey farmer, we only worked for the summer crop. There was very little honey to be had in the spring, and if there was we usually left it for the bees so that they could build up their strength for the main flow. These days we have just as much chance of a honey crop in the spring as we have in July. There have been some fundamental changes in agriculture over the last 20 or 30 years, and the swing from summer honey to spring honey is one of them. Another little realised fact is the continual lowering of the water table. In the whole of the Thames Valley area, boreholes have been sunk to take the water away for use in towns and industry. This has meant that more and more of our chalk streams are running lower and lower, and sooner or later they will not flow at all. I have seen the average level of the Kennet fall season by season, and we still pour the same amount of effluent into this diminishing volume so pollution levels are rising. This draining away of the underground resources means that much of the land is becoming dryer. Coupled with this is the ploughing and draining of the fields that were pasture. These fields no longer soak up the water like a sponge through springy turf, they now shed it like a raincoat and the winter rain, which should be there for the summer clover crop, is thrown straight down a drain and back into the sea before you can turn round. This sad state of affairs is getting worse year by year, and is affecting our summer honey crops, not to mention many other things.

From a honey farmer's point of view, these changes in the water table are very disturbing, and when they are coupled with the continuous spraying of our countryside, the problem gets even deeper. For me there are three main types of chemical spraying. First of all we have herbicides. These have decimated the vast areas of small wild flowers that filled our fields and water meadows just after the War. As I said, my great grandfather painted water colours and oil paintings of the Thames Valley about 120 years ago and the colours of flowers in his pictures were a riot. I have visited those same fields that he painted today, and all I see is a few buttercups. Herbicides have had the rest. In

Chiffchaff on Cherry Blossom

the 1930s and 40s, when R.O.B. Manley farmed the Chilterns for honey, he could put 30 or more hives in one place and still expect to get a crop of honey. Now, if I put more than a dozen hives in one place, they are overcrowded. Herbicides have destroyed the wealth of our summer crop.

The next killer, as far as the bee farmer is concerned, is the range of insecticides that are used. These, even if used rationally, still do unseen damage year after year and, because of this, the loss goes unnoticed. For example, the bumble bees of our area have been driven back to the roadside verges of our country lanes by insecticides and loss of habitat. The result for them is that, like a cat that lives by a road, hundreds of thousands of them are killed by cars spring and summer, and if one of them is the queen establishing her nest in April, then all is lost and nothing survives.

Lastly, we have fungicides. These are said to be safe for insects, but are they? I do not think so, in fact I am sure that they are not. The problem lies in the brood nest of our farmed and wild bees, and that is where the damage is done. What happens is the adult insect collects what she thinks is innocent water from sprayed crops and trees, and she takes home a potential poison to feed her young grubs in the nest. These grubs are delicate little creatures, and when they are fed on pollen and nectar laced with fungicides they die. This loss of young causes a decline in the family unit and the population can get out of phase, so causing a further decline in the unit's ability to sustain itself.

All these problems set out above will cause the honey farmer of today to struggle to get a summer crop of honey, and they are uppermost in our minds as we come into the spring, for this is the time of spraying. To me the countryside is an intricately balanced unit, it lives and breathes like any other living unit, and running a bee farm brings this all home to me far better than all the books I have read. All the time you can see the changes and you can see the damage that they are doing, falling water table, too much spraying, disappearing hedgerows, or hedgerows shorn like a service hair cut – they all point to fewer birds, bees, insects, ponds, and so no frogs or newts. I have watched the decline of the grey partridge, because its face no longer fits. As a boy as I wandered barefoot through the wild fields of Warfield, wondering at the flowers and butterflies, the young covies would burst from the grass at my feet, and glide away across the corn field to drop amongst the mass of red poppies at the far end. Today there are no grey partridges on the fields of East Berkshire, or if there are they are very few and far between.

This all makes for far greater planning of a bee farm in the spring. It becomes imperative to get the facts right, to read the season, to note the rainfall, to remember where it will be dry and where it will be wet. Where the wind will be at its keenest, and where the woods will give shelter.

We go in for detailed planning in early April. We try to plot out the progress of the hives through the coming months. First they will be on the fruit trees in the orchards and the oilseed rape fields. From there they will be moved to sites where they can reach field beans, and from there to their summer sites where there will be all the summer trees of lime and sweet chestnut. Also, the summer flowers like blackberry, clover, willow-herb and the hawksweeds.

This planning is thought out, many miles are driven to see farmers and see what they are growing, and discuss places where the hives can be put well out of harm's way. Bees are a help to the countryside, and with the loss of our bumble bees they are of even more use than before. New crops are being tried, and some of them need bees to get on well. Pollination of beans is a must, and so is clover, we have tried fields of lupins, we have had some success with mustard grown for seed. Sunflowers will do well with bees near at hand, and even the rape will do better with an early set of its first flowers.

Many springs ago, in the late forties, I became very friendly with the keeper in our part of the world. He taught me much of my knowledge of the countryside. As he stood cutting back a hedgerow and laying it on its side in the old-fashioned way, he would tell me about the little birds that lived in it. He would take me and show me the skylarks nest in the long grass, or direct me to the family of little owls in the hollow tree. In my spring holidays, when I was home from school, I learnt far more from that old countryman than I ever did sitting behind a desk reading the facts of the Trojan Wars in incomprehensible Latin.

One spring he pointed to the magpies building their huge domed nest in the thick thorn hedge, where the hawthorn buds were just tipped with green. "For every clutch of magpie's eggs you bring me", he said, "I will give you a shilling, those birds are villains to the nestlings of our warblers and finches". He smiled and drew on his pipe, "I am too old to climb those black thorns, but you aren't".

I found ten nests that spring and five of them were laid in. I watched them often, climbing to the nests to count the eggs. My old friend was not interested in one egg, it had to be the whole clutch for me to get my shilling. One morning I took seventeen eggs in a large white handkerchief for his inspection. He looked approvingly at the greeny-black collection, "I think that those came from four different nests", was his comment as he turned each one over in his hand, and of course he was right. "What are you going to do with them?" he queried, as he reached into his pocket, and gave me four shillings. I told him I was not sure. "Why not have them for your breakfast, you will find them as good as any other egg". I took them home to my mother and she scrambled them for me, and they were excellent and like the plover's eggs, the colour was a deep orange. Soon I was back at school with only memories of those free days, until my elbow became very swollen and stiff. I could hardly move it. Eventually I went to Matron; she looked at it and pronounced that it had gone septic, and after some probing she extracted a huge, long black thorn spike which must have been in there for quite four weeks. "What have you been up to", she asked, "You must have known about this thorn". I told her about the magpies and the large domed nest of thorn and clay beautifully lined with tiny soft roots. I could see that she was not impressed. "You should be working, Field, not climbing thorn bushes, think what it does to your clothes".

In recent years the magpies have done well for themselves. They have been one of the winners, in the modern urban communities that creep in and spoil our beautiful home counties. Their populations have risen season by season, and their familiar black and white form can be seen in many crowded back gardens and on most home-made bird tables. I remember one cunning magpie who, when he found the wood he was in was surrounded with guns out on a day's pheasant shoot, flew upwards in tight circles until he was well out of gun shot, then set his wings and glided to safety.

Chapter 5

May

Come with us and have a day working the bees in the orchards, the offer is always open but the work is hard and repetitive. There may be a hundred hives to be checked in the day, and we will want to get them finished before we set off for home. If the weather is fine and warm the bees will be of good temper and well behaved. On the other hand, do not come on a cold damp day with drizzle for they will see us coming and will come out to meet us! When we take the roof off the hive they will not be friendly and will need plenty of smoke if we are to control them. We will look for early swarming signs at this time of the year, and see that all the hives have plenty of food. There may be some that did not get enough winter stores and are now very short. There will be others that came from the willow beds in Dorset and have already had a honey flow. They may be getting a little too strong, and perhaps we had better take a frame of brood away from them and give it to that weak stock in the corner. Then there is that hive where we never found the queen on the first round. She will need checking and finding and, if need be, clipping. I don't know where she goes but somehow she seems to have her personal corner of the hive into which she dives just as soon as we lift the roof.

The next hive has a very old queen and she looks as if she is on her last legs. She will be no good and should be got rid of. That supersedure cell in the middle of the next comb is a good sign, we will leave her to get on with it. Once she has a new young queen we will give her some bees and build her up for the summer flow.

Come over here. These five stocks came up from Dorset, they all did well on the willows and all five will need a super for room, if not for honey, for there will be thousands of young bees hatching next week. The weather might turn hot in the next few days and then they will do very well off the cherries. I hear you ask about those pears over there – well that is where all that green pollen is coming from, but they never seem to get much nectar from pears. I have taken a super or two of honey from plums, and I have seen some good flows off cherries and apples, but never the pears. I even once got a flow off a field of gooseberries, and very good honey it was too, but they come out very early and the bees are not usually up to it.

Now if you look closely at this comb, you will see the problem that I was talking about a while back. Here you can see the fungicides have killed the grubs, they are standing up in their cells and have turned brown, they have not collapsed like they do with foul brood. I expect that during those hot days last week the apples were sprayed with fungicide and the bees looking for water just after we moved them into the orchards have collected the spray and fed it to the brood. It never seems to affect the bees themselves, though if they did die we might never see them.

We have now done nearly half the hives and it is approaching mid-day. It's time for me to stop for a cup of coffee, the sky has cleared and the sun is quite warm. It is developing into a perfect spring day. The hives all round us are becoming very active, the bees are streaming out and away and across the trees. In fact, if you look very

Cuckoo on Horse Chestnut bough

carefully, you will see that they are coming home with little yellow faces, that means they have found a nearby field of rape, and that will be far more attractive to them than the apples or pears, or even the cherries. Soon the whole working population of the hives will be out of the orchards and we will get some honey.

Sitting here watching the bees, listening to a distant cuckoo and the faint scent of the apples drifting in the air, who would go back to an office job, I ask you, and I have a whole summer ahead of me. But this won't do, we must get on and work through another batch of hives, looking at this one I really do believe there is a nectar flow; see how the drops of unripe nectar fall from the cells as you turn it over to examine it. The honey must be pouring in, this is the first real flow of the year, and each stock may take ten to twelve pounds in the afternoon. This unripe honey is full of water and would be no good if extracted, the result would be like milk. It will take the bees some three to four days to get the water down to an acceptable level. If the level of moisture is above 22%, then the honey may well ferment and a safe level would be below 20%.

Watch the bees as they fly home, see how they hang in the air with their abdomens low, that is a sure sign that they are full of nectar, their flight is slow and ponderous, quite unlike the way they zip from the entrance on their outward flight. Every bee in the hive is now so busy that they will take no notice of us, we will have a trouble-free afternoon. All the old aggressive bees are out in the field, and all the young bees have more than enough to do. Even the queen will be increasing her egg laying rate, and will have to search for empty cells to lay her eggs in.

We have nearly finished that load of fifty supers that we bought with us, and just in time I must say. Another couple of days and we would have been in real trouble. If you look closely you will see how all the empty cells in the brood nest are getting filled with honey, and if we did not put on these supers the whole hive would become completely congested within twenty four hours. As it is they will be able to take the nectar up into the super tonight, and so have plenty of room by the morning. Another day and the queen would have nowhere to lay her eggs, and in no time at all she would be filling up those little cups that are appearing at the edge of the combs.

Just look at your feet! Your boots are covered with wet honey, even your overalls are getting damp round the knees. This is the sort of day we see too few of, all we need is another three of them and we will have four or five tons of honey around the bee farm. The trouble with the British weather is its inability to make up its mind, and once a honey flow has started, some heavy rain will come along and put an end to it.

You see those five hives over there at the edge of the fir wood? Well they are the last, when we have checked them then we can go home. Nature is gathering speed and tomorrow will be another just like this. It will be a race against time to keep ahead of all the stocks. There are bound to be some stocks that will get too advanced with their swarming cycle, and we will have to resort to surgery on the next visit. These stocks will have to be split in two and the combs taken away will be used to boost up those little stocks that are not getting together. It is amazing what two combs of hatching brood will do for a hive that is only half way there.

Mid to end May are to me the two best weeks of the year. It is indeed what we call "clouded cuckoo time, when the showers betumble the chestnut spikes". It is good to be alive, all is fresh new green, the scent of the apple blossom is practically over, but the beautiful horse chestnut is coming into flower, while the hedgerows are splashed with the blooms of the hawthorn, with their heady aroma. As for the bees, often they are working the spring honey flow and building towards their best; both the horse chestnut and the hawthorn suit them, though the horse chestnut spikes are the ones that will fill our supers.

I would like you to come to our apiary at Thatcham on the Kennet. There is a stock there where we caged a queen and she needs checking. It has been a busy day and we have worked hard, but an evening by the Kennet in late May is not to be missed, the weather is set fair and the sky is still with a few wispy clouds drifting high along the horizon. We might as well put a rod in the back of the vehicle, for who knows – the mayfly may be about, and after we have checked the bees we could walk up the river.

To me there is no better time of the year to walk the river, there is so much to see and listen to. The nightingales will be back in the thick woods and we are sure to hear them. The swifts and swallows will be in abundance over the water's surface, taking the mayflies as they hatch. All you will hear is a tiny snip as they catch the insects, and turn to look for another.

The mayfly hatches one day, and if the weather is fine it will then shed its last coat to become that final perfect mature insect that dances in thousands all along the bank in the early evening. If you look you will see them in their masses in the shelter of the tall hawthorn

Trout taking a Mayfly

bushes. This is also trout time, when the giants of the deep holes and river bed come to the surface and gorge themselves on the feast.

When I first came to the Kennet some thirty years ago, this part of the river was full of these huge fellows, and during the mayfly season someone always caught one fish worthy of a glass case. Today the river has long since been spoilt, the mayfly is still there, but almost all the trout have gone, pulled out by coarse fishermen on worms and such.

I still love to walk the river, and on an occasion I may even cast a fly to a trout, but nowadays the river runs several feet lower than it did when my father and I first came to this stretch. Today the golden gravel over which the trout would hang and chase their food is too often above the very surface itself. We are told that pollution has not increased, and that no more outflow from sewage farms is allowed, but if the river has only half as much water in it as it had before, the dilution is halved. No fish can go on spawning and hope to survive if this continues.

Come, let us walk. I will not depress you further. Enjoy what is left, for as long as it lasts. You see that old iron bridge? We will walk down to it and rest our elbows on the rail, there are always many chubb feeding below it, and sometimes you will see a grand master of a fish. If you watch the mayflies drifting down, you will see the white lips of the chubb appear long before you can see the whole fish. Those lips stand out, and as you look the shape appears, floats to the surface and, with a gentle swirl, the fly is gone. There must be ten fish feeding away across the river. If you don't move they won't take fright. Many years ago some very fine perch lived under this bridge, but they have gone, they were killed by a disease that wiped out perch a good twenty years ago.

Did you see that rise just to the left of the willow branch that trails in the river? That was no chubb, I well bet it's a trout. Watch the spot and see if he does it again, there's a fly coming down, it's just about reached him. Just look at that great green shape with black spots, he must be three and a half pounds. He takes the fly with a savage snap of his jaws, unlike our gentle chubb. There you are – he has taken another. It brings back memories of long ago, when the river had many such fish and my father and I would return with a brace of fish weighing five pounds. It is sad how much we have lost, I think we will leave him alone. If he took our fly he would be fast in the willow roots before we knew what had happened.

Just listen to the warblers. This part of the country is full of them, the thick reed beds are their nesting sites year after year. The white throats are singing from the blackberry thicket, the willow warbler hop amongst the dense pollard heads. I can hear a blackcap somewhere in the distance and the chiff chaff has sung here since the 1st of April. As the sun goes down they all proclaim their presence to the world and their families.

As dusk gathers, the last house martin darts along the river picking up a dying mayfly as he hurries home to his roost. It is at this moment you will see the first bat appear; it is strange how one moment the swifts are screaming across the fields, and the next all is silent with the first movements of the creatures of the night. In years gone by there were many bats to be seen along the river at this hour. In the half light it was difficult to see what they were, but today there are far fewer. One or two long eared and a number of pipistrelles, but I have not seen any of the larger species for five years or so. The river settles for the night. There are still owls along the bank and a tawny appears in the crown of a willow, it is time we were off home, we have a couple of dozen hives to move tomorrow morning, and bees start flying rather early in late May. We must be at the orchards by six, or we might be too late, and they have said that they wish to spray the apples for coddling moth. Listen, the nightingale and the sedge warbler are still singing.

By the end of May we seem to be at a peak of activity. There are hives to be moved and there is often a crop of honey to be taken, we usually have a ton or two off the hives by the end of the month, and of course the bees may well be preparing to swarm. It means a very long day, and too often a conflict of jobs to be done. Do we get on with the swarm control, and making up our spring increase, or do we work hard to get the spring rape honey home and into barrels before it sets up in the combs. At least in late May we have plenty of daylight and so plenty of hours in which to work, the only drawback is the fact that bees usually fly during daylight, and unless it starts to rain there will be too many of them on the wing for us to be able to shut them up and move them to their new sites. This will necessitate a very early start, or a very late return to our beds. Only last summer we went to move some bees away from field beans which were to be sprayed. We left after an evening meal, picked up the hives by 9.30, went straight to the bell heather near Poole Harbour, were unloading bees in hives with two supers on each by midnight, and on our way home by 2.00, to be in bed as it got light again.

Bees will often be flying by 6.30 in the morning, and on bright warm days we have to get to them by 6.00. It is a lovely time to be out, and most of the roads that we use are very quiet. Rabbits hop along the verges, the dawn chorus is just finishing, pigeons flop across the fields as if there were no men left in the world, the carrion crow pecks at a dead hare by the roadside, and a roebuck stand peacefully at the edge of the wood taking the last of his meal. In the orchards we hurry to the hives to make sure the bees are not yet active. We pull a tuft of dewy grass from the side of the hive and wedge the entrance block into place. A few bees may fly out and sting ungloved hands, but once the entrances are shut we can relax and start to get the hives loaded onto the vehicle. A screen will have been placed over the top super the day before and the roof replaced, this will give the bees the ventilation that they need for the journey. From the orchards they will very likely be going to field beans, or clover, or to a summer site; there may be a late field of rape to cover, or even a field of lupins.

Chapter 6

June

As we swing into early June, we are approaching the top of high tide, the season is half over and there are only two more weeks before the longest day. Growth in the countryside is at its peak, all the greens are at their best, the leaves and the grasses are still all young and full of moisture. The buds on the lime trees will still grow a little before they flood the avenues with scent and proclaim mid summer to the world. In the woods and chalk banks this is the time of the spotted orchid, there are still areas where I can find plenty of them. Where the new motorways are banked up, the seed that has been dormant for tens of years will be thrown to the surface and a mass of flowers will suddenly appear. I have also seen pyramid orchids later on in the summer growing on a bank for the same reason. These lovely flowers have become very scarce and have to be searched for. In Dorset, when we take the first load of bees down to the bell heather, we always go to a favourite marsh in the Poole Harbour basin to see the marsh orchid, they grow along the roadside in the soft spongy soil and stand nearly a foot high. When I was a boy, there were fragrant orchids in a field near Bracknell, and every summer I would go and pick my mother a little bunch. The field was dotted with little hawthorn bushes and had never been ploughed for fifty years, it was used for cows who would wander amongst the bushes. After the War the bushes were pulled out and the field ploughed and planted up to wheat, and I never saw another orchid.

Spotted Orchids

There is much to see and every day in the summer we get our surprise. Last summer it was a pure white spotted orchid, the first I have ever seen. One early May day we saw an osprey flying up the Thames. She came straight over us and there was no mistake, no doubt on her way north to join her mate for the summer.

For those of us who like to spend the short nights of early June away from our beds, those hours are full of activity. If you go to the bees on a warm night you will hear the gentle hum of fanning bees as they evaporate the moisture from the nectar they have gathered to turn it into real honey. If you are lucky, you may see a hawk moth go whispering by on ghostly wings as he searches for a mate. This is the time for these giant moths, but they are very secretive. While I was at school in Dorset a friend of mine used to put out a moth trap in the summer term and we caught at least six different types, including two deaths heads, but I feel there are not as many as there were. The other night wanderer who has become very scarce is the stag beetle. There was a time before the coming of Dutch Elm Disease when I could go out any night and catch a dozen, but today they are few and far between. Their problem is that the elm tree was the host to the stag beetle grub. This grub lived in the elm tree for six or seven years before it hatched out to spend one summer among the elm's branches, flying from tree to tree at dusk. Now there are no elm trees to speak of, and those that we have are all very young and hardly fit to house a population of stag beetles. I rather feel that we may have to wait a couple of hundred years for the trees to become big enough.

The field beans are a Mecca for bumble bees and, if we are to see them, this is a crop that they love. It is they who, in dry years, nibble the base of the bean flower to get at the nectar, and so make a hole which is very useful to the honey bee, who would have trouble in reaching the nectar from the flower itself. When the weather is hot, wet and sultry, then there will be plenty of nectar for all, but in dry dusty seasons the bumbles do us a service.

On the bee farm in early June, all those strong stocks that have not been properly catered for may well be found to have swarmed, or be about to swarm. This is an ideal time for colonies to split up, otherwise, if they get the chance to swarm they will. It is my opinion that like most problems in life the sooner they are dealt with the better, and if we can weaken down all our very strong hives in early May, and give the extra bees and brood to the weaker hives, then swarming at the end of the month will be far less. It is no good waiting until the problem is upon you, and then trying to do something about it. In our book the problem should not be there in the first place, and once the bees have sealed queen cells in their hive there is very little you can do about it. Besides there will be so much other work to do in early June we will not want to be wasting our time trying to find virgin queens, or splitting up good stocks, which even if they mate, a new queen will not be very much good for the rest of the season. Swarming is a natural instinct and must be suppressed as far as possible and not frustrated. Suppression may still leave you with a honey crop, whereas frustration certainly will not.

It is in June that field beans come into there own, and if you are lucky you will get some first class nut brown honey with a fine flavour. Field beans have become more and more popular in recent years, and we put bees onto many hundred of acres of this crop. When there is a flow in hot humid conditions, the honey will pour into the supers, and the bees that were quite well under control will suddenly boil over with maybe fifty percent of them trying to swarm. We used to speak of the June gap in bee farming, this was the month between the spring and the summer honey flows, but nowadays with modern farming, growing a June flowering rape and field beans, there may well be honey about right through the month. In fact these days there may well be honey about right through May, June and July.

June is the month when many farmers will be spraying fungicides on to wheat crops, and quite often we are asked if this will kill our bees. I always say that as far as I know, it will do no harm to the bees, as they will not be working wheat fields, but if they did collect from a sprayed field, the moisture so created may well kill the brood. What it will do is to kill the spore of fungi in the fields, and these fungi do a useful job eating up the rotting vegetation of the countryside. It is another case of interference in the natural chain of things, and can surely do no good in the end.

Many years ago mushrooms were very common from July to October, and only disappeared with the first frosts. Nowadays in the cultivated parts of the country, mushrooms have almost gone, it is only down in Dorset when we are working the bees on the heather that we still see field mushrooms in abundance.

A few years ago we had an apiary at Ibstone, in the Chilterns, in fact we still have it, when I first worked it some twenty years ago the field around it had never been ploughed, it was rough grazing, and from late May until September there were often huge horse mushrooms to be picked. When I arrived there with Harry Wickens we would look around for them, I once took one home which weighed just on half a pound, and was as large as a dinner plate. This ring of horse mushrooms extended over quite an area, and every year they would come up when the weather was warm and wet. Then one year the farmer ploughed the field, pulled up the hedge and added this bit of rough grazing to a much bigger field. Never again did we see the horse mushrooms, and for a dozen years I thought they had gone, then last season, right at the back of our apiary, where no plough or sprays have been, my son found a small horse mushroom. Perhaps the ring may grow again even if only between the hives.

Harry Wickens once told me a lovely story about fungi. When he was a young man in the late thirties he was working with R.O.B. Manley and they came to an apiary out in the chalk hills of the Chilterns. There growing against the hedge were some Morels, a rather spongy looking fungi, quite unlike any mushroom.

"Ah" said Robert Manley, "Morels, I love Morels" and he picked himself a hat full, "I will take these home for my tea". Harry said that he had grave doubts that he would ever see the old man again, so when he appeared the next morning, Harry asked "How were the Morels?" The old man looked up, "Oh, they were fine, they're rather like beef steak".

A week or so later they were again working the bees at the same site, and there by the hedge was another crop of Morels. Robert Manley hopped out of the vehicle and walked across to gather them, he came back with a hand full, "There you are Harry, take those home to your landlady and get her to cook them for you".

You can imagine what she said to Harry when he suggested that he was going to eat these dreadful looking toadstools for his tea. She took a lot of persuading that there would be no need of a doctor at some time during the night. But, as Harry said he had to eat them, R.O.B. was bound to ask him what he thought of them, and if he said he had thrown them away the old man would have been most upset. As things turned out he said they were very good, and his landlady could hardly believe her eyes when he came down to breakfast as right as rain. There are of course only a very few of the mushroom and toadstool family that can really do us any harm, the trouble is that the ones that can can more than damage your health, so we leave many good things alone, for fear of making that one fatal mistake. Harry knew the mushrooms that we could eat, and directed me to several that my mother would have had me wash my hands even if I touched them.

As I have said June is one of the busiest months of the year for the bee farmer, and even when a moment does arise when we can stop and think for a second, a farmer will ring up to tell us that he is about to spray a certain field, or ask if someone could call in and see him about pollinating the beans that he planted that spring. He has heard that bees on beans will give a bigger crop, and as this is the first time that he has grown the crop, he wishes to know more. Much as we like this type of call, especially if a fee can be arranged for the pollination, it does keep us on our toes, when we might like to sit by a river bank instead.

I have had just such a call, and as I have to go out and have a look I thought that you might like to come along too. The field is at the back of Marsh Lock, and a walk down the tow-path will be pleasant. We have got to go and find a corner to put some bees, well away from any public foot paths, and where they can easily reach the crop. These spring sown beans flower much later than the autumn planting, and come out from mid to late June.

The question of placing the hives is very important, and attention must be paid to the public or you may find yourself in trouble. A few years ago I placed fifteen hives on a field of these spring beans somewhere near Harwell. The results were rather horrific. Behind the field there ran a rough gravel track up onto the Downs, on my several visits I never saw anyone use it. I decided to put the hives on the field behind a thick low hedge that bordered the road. On the Saturday morning I went over and did a full swarm control on all the hives, they were fifteen of our strongest stocks, and when I left them the air was thick with very angry bees. They had got what you might call "out of hand", I left in a hurry, with not a soul in sight.

What I did not know was that a certain race horse trainer took his twenty steeds up that very track on their daily exercise. They went by about half an hour after I had left, and as everybody knows, horses and bees do not get on very well. The result I understand was a stampede, in fact the words from the race horse owner to me on the telephone later that day were practically red hot. I have never known anybody quite so angry with me, well not since I dropped a pass in a house match at school. "Do you realise", he thundered, "Two of those horses are running at Ascot next week". I was tempted to say that I would put a pound on each of them, as they might gallop a little faster, but I do not think that he would have been amused, and might have appealed to a higher authority. I pointed out that the law says that bees are wild creatures and did not belong to me once they had left the hive. A few days later I took him several pots of honey and said how sorry I was, but he was still not best pleased.

The lesson is, to be very careful indeed where you site your bees, and look for any possible complication. The problem in this case was that I had just looked through these stocks, and by the next morning the horses might not have seen a bee.

If we walk down this path it will take us to the river, we can leave the car here, it is a pleasant evening and there is half a mile of very attractive water to walk along. The green in the trees are at their best, though I must say I miss the great old elms that used to stand over the other side of the fields.

Many years ago I came along here with my father on just such an evening, I remember it well, it was the 25th June. We worked together in the family business, it was long before I became a bee farmer, and on that evening we had decided that as the fishing season had just started we would walk down to Marsh Lock and fish off the weir in the weir pool. My father knew old Jim Tame, the lock keeper, and he was expecting us. The sun was getting back over the trees towards Henley when we settled ourselves down by the open sluice gate, the soft smell of river and damp weed drifted up our nostrils, an evocative smell. Soon my father had a tackle with a large worm on the hook, right out in the middle of the pool, we sat and waited, we talked, my father smoked his pipe, and the sun went down. Suddenly the rod top dipped dramatically, my father stood up and I could see that he had hooked a large fish. As the time passed we realised that it was a very large fish, but we had no idea what it was. After about twenty minutes a huge olive shape appeared at the edge of the weir race, for a moment I saw a great fish roll over and then it dived back through the clouds of

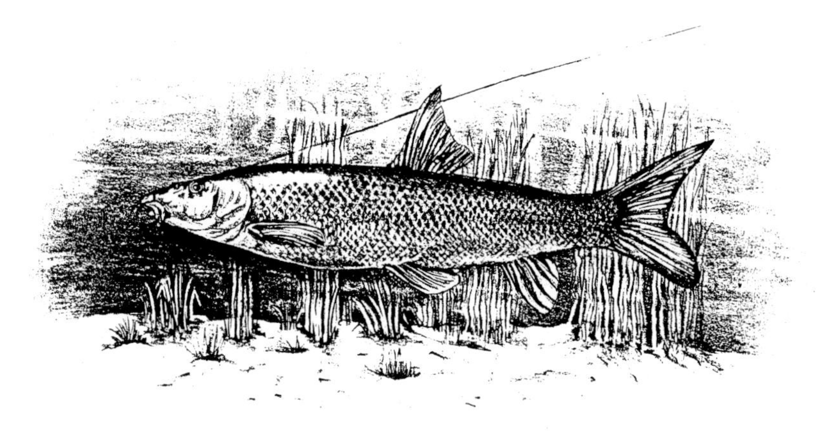

Thames Barbel

bubbles to the bottom of the pool. The lock keeper had seen the struggle and hurried along the top of the weir with a massive landing net, "I think you have hooked one of the big Barble", he said as he reached our side. Moments later a great head appeared on the surface and we could see he was right. A few more minutes and the big fish slid into the folds of the net, without that net we would never have landed him. We took him ashore and my father gently hung the fish on his spring balance, it plummeted to its limit of ten pounds, it was probably nearer eleven, that was the opinion of Jim Tame. We put the fish back into the net and lowered it back into the weir pool. It hung in the still water for a few seconds and then with gentle strokes of its huge tail it swam back into the depths and home.

That was all of thirty years ago, but this part of the river is still a lovely place to walk in the evening, in fact in years gone by, I picked, and you must forgive me, for I was very young, fritilaries on this very field and took them home to my mother. The beans are across that little stream, on the meadow the other side, I can smell them from here, looking at the situation I do believe that our bees on the Wargrave-Henley road, on the other side of the river, will reach this field. It is no great distance, not much over half a mile, and with all this scent they will soon find them. I think that we can safely tell the farmer that he will get his pollination, and we won't have to do anything.

By Mid Summers day our bees should be reached their greatest population, the cycle that builds bees from early honey and pollen, to gather even more honey and pollen to give us even more bees, will have reached its peak, and just like the passing of high tide on Mid Summers day, from now on the numbers of bees will decline. This decline will be tempered with the strength of the summer honey flow in July, the heavier the flow, the faster the colonies will decline. Like high water, little will change for a week or two, but to me once we are passed the 1st July, time is running out. That peak of population will coincide with the flowering of the lime trees and that heavy scent of the sweet chestnuts. All around the summer flowers will come out, the willow herb, the white clover, the blackberry, the thistle and knapweed. There will, on hot days, be a surge of activity, and a mass of bees will start the summer work, supers will begin to be filled, and at that moment of high tide, those that have not swarmed already, and feel the need, will be away, we must be vigilant. One week the kettle is gently singing, all is well with the out apiaries, the next we are over the top of the hill, the summer flow is spread out before us, there are flowers in the hedgerows, along the river banks, in the great trees, and across the meadows. This is what all the work has been for, one last effort on swarm control, and the season will be ours. Summer flows are never as dramatic as the spring flows, but they seem to be far more sustainable once they get going. In the early seventies the summer flow stretched for three weeks, with a steady intake of honey day by day. In the spring it is often all over in a flash, we may have four tons of honey in as many days, and then the weather will break, in the summer the flow gets under way and slowly builds. Gradually the countryside dries out and the last drop of nectar is sucked from the earth. In other years it may just have got going when the weather will turn, and cloud and wet will put pay to a year of promise. When this happens we will look into our hives and see them overflowing with unused bees, and we must make preparation for the heather, so as to use up this valuable stock, which would otherwise go to waste.

Chapter 7

July

Why not come with us to Dorset at mid-summer, we have got the first load of bees to the bell heather at the beginning of June, but the second load will go down about mid-summer, the bees have been on spring sown field beans, and they have only just finished flowering. The farmer wishes to spray them with some lethal concoction to kill the black fly that has started to appear at the tops of the plants, he tells us that it has been too dry for a systemic treatment, and if he does not act it will be too late. He would like the bees out by tomorrow, so perhaps we can move them tonight!

We look at our watches, it is mid-summer, and the weather is warm, it is just as well that we put the travelling screens on last week when we took the bean honey. The bees will not stop flying before nine at the earliest, and it will be eleven before we can expect to get the two sites cleared. Some of the hives have two supers on them, and with the travelling screen they will be quite a job to handle. We can get sixteen supered stocks on the trailer, and may be another dozen on the vehicle. Thirty stocks will be our maximum, we will try and pick that number up, we can leave those two little stocks behind, they will be of no use to any one, they swarmed a week or two ago.

I had better make a thermos of coffee, I can't work without coffee, especially in the middle of the night. We should reach the back of Poole harbour about two thirty, and get the bees off in under an hour. By the time we have unloaded and I have had my coffee, the first shadows of dawn will be creeping across the sky, the Nightjar will have stopped his nightly chirring to his mate, and we will hear the call of a wake-full Curlew.

Unloading bees in the dark is not without its problems, and on this night one or two staples have moved, and in one case a super has been pushed off the brood chamber. It is of course the first hive that we come to move, and it is covered with bees, they are crawling everywhere. In the dark they will not take flight, but they have clustered on several other hives. It is time to put on a pair of gloves and hope they don't get up your sleeve. I once had to move a hundred hives back from Kent, which I had banded with a banding machine, and not stapled, on this night the weather had been very dry for a couple of weeks, and as soon as I picked up the hives the supers and floors slid away in all directions. I have never had so many bees out on the loose, I was stung some two hundred times before I had got the hives all loaded and roped up for a safe journey home. They crawled to every part of my body, and stung me every time I moved. It was a true nightmare in reality, and it was only the songs of several Nightingales that kept me from giving up and sitting down in total desperation. On the way home we stopped at a pub in Windsor Great Park for a pint of beer, about mid-morning, and left the lorry parked outside, the bees were still getting out. After we had quenched our thirst we went out into the car park, and you have never seen such a sight, there were bees flying in every direction, and some of them were coming back with pollen on their legs. As we left the landlord came out and passed some remark that I did not catch in full, but I made a mental note not to go back to see what he would do with the pint that I might order.

On this occasion we lift the last hive off with relief, and stumble through the rough heather stems, to place it on the last space on the stands. We always like to have stands for bees on the heather, it means that they can stay there right through the winter if need be, and not be at any risk. Coffee is my next priority, and as I pour out the flask I hear the first skylark launch into song, he has been singing since the spring and will soon be silent. There will be too much work with the nestlings in the heather, and by late June the season of song comes to an end.

After Mid Summers day the season seems to drift for several weeks. We have reached the high water mark with our colonies, their populations are at their peak, and all is ready for the main honey flow. Usually this comes in our part of the world some time between the 1st and 12th July. June is a very busy month without a moment to relax, but by the end of that month everything should be in its place, the supers on the hives, the swarming well under control, and the main moving completed for the moment. If ever there is a time in the summer when we can take a day or two off, and take stock of the world, it will be in July, and it won't be for very long, but in most summers I find time to break with the work.

During my life I have been asked "Name your favourite luxury" and without fail I always reply "a hammock". There is nothing I would rather do late on a July afternoon, after several hours out working with the bees, than to go up into my plum orchard and stretch out in a hammock between two of the old trees. First I will take a large real sponge soaked in cool water, and squeeze it over my head and dry my face of sweat, then I will pour myself a pint glass of lime juice, add a few ice cubes for good measure, and retire to my hammock.

To me this is pure luxury, to kick off my shoes and relax for the next hour or so. In July this is possible, and there are occasions when the weather will be kind, and the hours will pass and you will still find me there. The trees are in full leaf, the young plums are still very green, but well formed, the trees are full of young birds, the white throats chitt to each other as they look down on me, the Blue Tit families hurry from branch to branch searching for green caterpillars. In past years the purr of the turtle dove soothed the passing of that hour, today they have nearly all gone, but I now have the company of the collared dove instead. It is amazing how much you see just doing nothing. In these days the world is such a busy place, that no-one ever stops to look around, they rush and rush, and I often wonder if they ever get anywhere at all. Through the branches of the trees I can watch the House Martins and Swallows coursing across the sky, being outstripped by their brothers, the Swifts, who never stop screaming at each other. I have seen my first Swift as early as the 1st May, and yet they have all gone by the middle of August, I saw just one bird on the 12th September last year, and this was very late. If ever there was a bird that is the epitome of the British summer, it must be the swift, he only arrives when the insect population is about to explode, and he is gone long before the evenings start to close in. He is the bird of Mayflies, of Blue Winged Olives, of a thousand chiromoids, and dare I say the odd queen bee.

On occasions I do not get to the hammock until it is evening, but in July it is possible to enjoy the hour of dusk, and watch the shadows gather. By July most birds have stopped singing, they are too busy with their broods, but the odd Thrush still adds a note to the dusk, before the bats appear. As I have said it is sad how many fewer bats there are these days, and we have to wonder why, I believe that with all the spraying of our fields, this has caused, first a loss of food plants for our summer moths, which in turn will very much reduce the population of moths. Bats live off these moths, and if they are not there then we can support fewer bats. Chemical treatments of old buildings have not helped either, or the cutting down of old and rotten trees. As a prep-school boy during the last war, I was sent to a huge country

Frog and Water-Lily

house in Somerset which was surrounded by out-buildings. One July morning I was up very early with a friend and we watched the bats return, there were hundreds of them, crawling up the walls and under the eaves of the clock tower. I have never seen so many bats since, and often wonder if there are any left in the clock tower these days. At this school we collected the caterpillars of many moths and butterflies, and I well remember collecting a jar full of drinker moth caterpillars on a wet July evening, the friend who collected them with me happened to rub his eyes and came up in the most dreadful rash, we all thought that he was going down with something lethal, but he survived and I saw him only last year, so it did him no harm!

There is a pond at the edge of my orchard, and I have encouraged frogs and toads for quite a few years, they come out of the pond in that dusky hour while I lie in my hammock and start to croak, and walk round the lawn I have to be careful where I put my feet, and can count them as they search the grass for slugs and other creatures. Believe it or not I get to know them all by their markings, and I once had one that lived for ten years. Once frogs become mature and fully grown, they seem to be able to look after themselves better and better. Very few tadpoles survive in most ponds if there are fish in them, and small frogs and toads fall easy prey to birds, I have seen a Blackbird hunting along the edge of my pond and swallowing the little frogs whole. But once the frogs grow up, and this may take three years, they seem to live for many more years, often sitting on the same rock or lily leaf night after night.

Eventually it is too dark to see anything and the shadows of dusk reach out into the night for the shadows of dawn, an owl hoots in the distance. I role off the hammock and walk across to have a last look at the breeder hive in the corner of the orchard, at this hour you will be able to tell if the main honey flow has started. Crouch down by the entrance and listen, if there has been a honey flow there will be a steady draft wafting out of one side of the entrance, and a inward flow on the other. The bees will be evaporating the water from their nectar and passing the damp laden air out to replace it with dry.

Some time in the month of July there will be a honey flow, whether it will be large or small is in the lap of the gods, all too often it starts off well, and then dwindles away in cold damp cloudy days that seem to go on for weeks. In the last twenty years I have never known a complete failure of what we call the main flow, but I have known some years that have been all but a disaster. As I have said the flow will generally start at the beginning of July, but some years I have waited to well after the middle of the month for the weather to break. July seems to be a rather wet month in our part of the world, and it is only in years that are hot and dry that we get a real crop.

Look around you in July, for July is the month of summer flower, all will help our income. Blackberries hang along the hedgerows, they used to be far more widespread, but today with modern cutting equipment, our hedges are often ruined long before the flowers of the hedgerow have a chance to open. There will be patches of willow herb in the corners of fields and in the clearings in the woodland, though not native to our land it is a great pollen and honey yielder. Don't forget the thistle, he is a lovely plant, and should be encouraged, he has a deep tap root which can seek out moisture even when it is a little dry and other summer flowers have dried out. I have a good farming friend who makes a habit of kicking out thistles when he sees them in his fields, I once asked him why he wished to destroy such a lovely honey plant. He looked puzzled for a moment, and then with a smile he said, "Oliver, I am not going to start growing thistles for you, or anyone else".

In days gone by, our wheat fields would be full of poppies, cornflowers, marigolds and knapweed, all would contribute to the crop. Even now when the bees are near a field that has not been sprayed, and is full of poppies the combs will become choked with coal black poppy pollen, the bees seem to love it, though poppies as a nectar plant are not of great use. Summer days pass the bees are happy and very busy, the huge numbers of late June will start to decline, and by sheer hard work our population dies away, to leave solid slabs of honey in their place for the next generation.

One of our great summer losses has been the decline of the harvest mouse, who would wind his home between the stems of the standing corn, and peep from his woven nest at the prying face from the world outside. In years gone by, the harvest mouse was common, he was often collected with the sheaves of corn and laid to rest with the harvest in the row or ricks in the farm yard. There he would stay, and even breed until the thrashing machine would come chugging down the lane to start a week of frantic work.

As boys we would collect in the yard for the thrashing, and watch the sheaves as they were pitched by fork into the thrasher. Mice would be seen climbing down the sides of the ricks as they were slowly pitch forked away. Many sandy little field mice, with their beautiful white fur along their bellies would drop and run among the multitude of common house mice. To me it is small wonder that we have lost most of our population of barn owls. In those far off days that I remember, there was an ever present larder set before the barns, and this larder could support quite a large family of barn owls through the bleak months of winter. They only had to sit and watch the ricks to see mice scurrying to and fro, even the young birds of the last summer could be well fed until the thrashing, and even after that there would be many mice that had escaped into, and under the barns. Today the food never reaches the yard for the mice to find it, and as the supply of ricks and mice has gone, so have our barn owls. I still see them quartering the rough ground along river courses, and watching wild patches that have not been cultivated, but the free meals of the long winter months have gone, and too many young birds never reach their second summer. In those days when the owls were still with us, my parents would talk of the barn owls snoring in the bell tower at the local church, one night I was taken to listen to them, and my mother said that they were almost as noisy as my father, which he fervently denied.

Towards the end of July the countryside will start to dry right out, and most of our summer plants will cease to yield, seeds will be set and stems will harden and turn brown, the flow of honey will turn to a trickle, and the bees that are left will become very spiteful. These will be the last of the old bees, and with nothing to do they will tend to sit at the entrance of their home and chase off any intruder.

In the latter half of the nineteen forties, a major summer crop was the red clover, and in late July and early August many fields were carpeted with the deep crimson flowers. In those days I would return from school in the latter half of July, and at the same time these fields were at their best. As a honey crop, they were considered to be useless, there was a long-held belief that the flowers were too long for the bees to reach the nectar. It was the honey farmers like R.O.B. Manley and David Rowse who proved this not to be the case. In fact, David Rowse told me that in the summer of 1960 honey poured into his hives from this source. He said that the bees were filling a super a week and that some stocks filled three supers.

I have seldom gathered red clover honey, as in these days the crop has gone out of fashion. In days gone by, when the horse was still a main means of farming, and in the war when fuel was short, much transport and tilling was still done by the cart horses. These great animals needed hugh quantities of hay to feed them through the winter. Red clover hay was one of the prime sources, and the other fine hay was sainfoin. Both these flowers are of the pea family, and both of them were great honey yielders. There was the first cut towards the end of June, and then there was a second cut in August. With the demise of the working horse, these two lovely sources of nectar disappeared and we no longer had huge pink patches on our landscape. For myself, I believe that in dry years it is very difficult for the honey bees to reach the nectar in red clover, but come a damp summer with plenty of wet spells between the sunshine, then the red clover will yield very well, and as long as the weather lasts, so will the honey flow. All clover crops tend to be fickle, and the relation between moisture, humidity, heat and available bees is very important. I well remember taking a load of bees off a red clover field where they were collecting nectar in early August and taking them to Dartmoor, where they ate up all the clover honey and returned starving.

Come back with me to those glorious summer days just after the war. Watch the horses cutting a swathe round and round the clover field, see the meadow browns, the gate keepers, the blues, and even the odd clouded yellow rise from the field as the horses pass. It was a

Mecca for butterflies, they must have come from miles around, following the gentle scent across the fields. Then there were the bumble bees hurrying through the heads. It was a surge of activity while it lasted. Not only was the hay a useful product to feed the horses, but the clover acted as a nitrogen fixer for the field, and red clover every third or fourth year was said to do much good. As a boy I would wade through the crop. It would come up to my waist and when the florets were sucked you could taste the nectar.

The little Bull Brook mentioned earlier wandered through those distant fields. Often I would walk its course along a footpath that bordered the clover field. I would take a rod and a two pound jam jar and sit under the alders to fish for minnows and small dace in the heat of the afternoon. It was high summer, no-one would find me there, it was peaceful and quiet, a kingfisher would flash by, the yellow water lilies would sway in the current. The stream was clean and pure, full of small fish, even under the stones on its bed there lived hundreds of stone loaches, and further up its course I had located a great family of swan mussels. It was along this lovely stream that I met my first colony of bees, a truly wild colony living in a pollard willow. There was a slit in the trunk near the crown and, as I passed, I could see much activity. It was a hot day in July and there was surely a honey flow. I stood and watched the hurrying insects. I knew nothing about bees, except for the fact that they stung and produced honey. Here were bees as nature had intended, living their own lives in an unspoilt Berkshire hedgerow. For many years they were there, and often as I walked along the path I would look up and see them. Then one day I noticed that the willow had been scorched with fire and the bees were gone. Someone had tried to get them out and no doubt take their stores.

It was in those same fields that I found my first hornets' nest. There was a giant popular standing all on its own in the middle of one of the meadows. At home, several hornets were seen around the house, and my mother told me to leave them strictly alone. She said that one sting would kill me. One afternoon I watched the direction that one of these creatures took when I let it out of a window up which it was coursing. As I stood watching, I saw another hornet high in the sky making for the tree. I approached, and lo and behold there they all were flying in and out of a great gash in the side of the tree, about ten feet up. I left well alone, but in the winter holidays I climbed the tree and found the old abandoned nest hanging in the hollow of the tree. At that time of the year there was no risk, and I tried to lever the nest out with a stick, but it broke into pieces.

Go to the Bull Brook today and you will find a disaster. Gone is the lovely clean stream, the water is polluted, strings of silk weed hang from the rocks and stones, the fish have long since gone. Plastic cups sit in the mud beneath the bridge, paper hangs on the trailing branches, a wood pigeon flies down for a drink; today he does not seem so keen. A broken, stained swan mussel shell glints in the depths of a pool, but of the rest of his family there is no trace. I don't think we will bother to come here again.

Chapter 8

August

As a young man in my late teens and early twenties, I spent many happy weeks wandering around Dartmoor. The moor can be very beautiful, and it can be quite threatening. Some thirty-five years ago I would walk miles along the Dartmoor streams casting a wet fly across the fast flowing stickles and through the deeper pools. The trout were small, but there were many of them, a dozen a day could be expected, and my father who was better at the task would bring far more to the bank. It gave me a chance to get to know the moor, so that when I became a bee farmer, I knew where all the best heather was and in some cases who the farmer happened to be. In 1972 I took my first lorry load of bees to the moor, we put them down near Chagford, all in one place, some sixty six hives. They were M.D. hives, which have a large brood chamber, and that year the weather was poor, there was a small heather flow and we only got about five hundred weight of honey from the lot, although the brood chambers were stacked with food for the winter.

There were better years to follow, and in 1975 after the first long hot summer, we once again took sixty stocks. The roads were not half as good as they are today, and it took us five hours to get the bees from the Chilterns to the moor, not to mention the loading up before we left, and of course the unloading when we arrived. For two men it was a fine twelve hours work, with very little rest, about an hour between arriving and it getting light and starting to unload. This year the bees were set on the hill above Wydicombe, and the farmer was named Mr. Northmoor, so was the moor on which we put the bees, so his family had been there for a very long time.

The bees flew across his two top grazing fields, and then they were amongst the heather. The sun shone and it was warm, for a month the bees were busy, and when it came time to collect, I left the night before so that I could be with the bees at dawn, and get them shut up, the others came down in the lorry over night, and I was ready for them. There was a grand crop that year, and many of the stocks had solid supers of ling honey on them. I took a kettle of water from the stream in the valley and had a cup of tea ready for the work party when they arrived, the farmer gave me a pint of milk. That was the way of heather in the seventies. I lay on the top of the dry stone wall, all the hives were shut and I watched six buzzards climb in gentle circles as the sun rose and the moor warmed up. The sky was empty, hardly a cloud in view, the birds got smaller and smaller as they drifted upwards and upwards into the blue depth above. A spider came out from under the rock on which my chin was resting and started to stalk a fly that had settled on a warm patch of the rock that was in the sun. Gently he crept across the rock, keeping close to the surface until he thought he was close enough, then a quick dash and he had his breakfast. Time passed and I saw the lorry come over the distant Haytor hill, twenty minutes and they would be with me, time to put the kettle back on the calor gas. The bees were starting to complain, they had found out that they were not going to be able to fly today, they clustered on their screens as the sun got hotter and fanned, I had already taken off their roofs, and placed the hives in the roofs, a fine way to travel bees. Soon we were working and sweating, it took much lifting and stacking and carrying. The fickle Dartmoor weather decided to take a hand, dark clouds started to drift over the western hill and spots of rain dashed on to the dry rocks staining the lichen. A heavy shower arrived and then another, by the time the bees were all loaded and roped, the ground was getting very wet indeed, time to go, but would the moor let us go, the wheels started to spin and slide. We were getting a little fed up, we might have to unload the whole lot again to get the lorry off the moor. But then the wheels gripped and we were away, a close run thing, but we had made it.

I looked back as I drove off after the lorry, and saw a buzzard drifting low along the skyline, he swung up over the rocks and settled in the top of a stunted thorn bush, he was home.

I still love Dartmoor in all its moods, I have taken quite a few tons of honey from it, but beware of these crops from the high moors, they may have rather too much water in them, and if you don't keep an eye on them they may explode in the later months, forcing off the lids of the honey tins, and foaming down the sides. A little heating at the right time should kill the yeasts.

A last thought before we leave the moor, take a look over there into the valley beyond Haytor as we drive away. Down there is a little stream, and when I was still a boy I went there to fish. The little stream has been dammed, and there is a little lake about a hundred yards long surrounded by thorn bushes. It is quite deep and full of Dartmoor trout. One April evening I went there with a little spinning rod, and I spun the lake with a fly soon, I caught several little trout up to about half a pound, then I had a take which was the grandfather of the lake. A huge fish of at least three pounds, he took almost all the line that I had, and he jumped, a black trout with golden sides and red spots. He came back towards me and got round a rock deep down in the peaty water. I saw him roll over and out came the hook. I think that it is time we left Dartmoor!

Now in August our thoughts are concentrated on the Ling Heather crop, in bee-keeping terms the heather flow has always been an after thought. We get our spring honey and hopefully we then get our summer honey, and many bee-keepers leave it at that. For me this should never be the case, for here is a lovely crop that fetches a high price. Anyway, what better way to spend our time in August than on the moors, when they are at their best and the purple expanses shine in those warm summer days.

Why not come for a day on the Purbeck peninsula, why not enjoy that warm sun, there will not be too much work to do, we will check the bees during the morning, just to make sure that there has not been a heavy flow, and put on the odd honey super or two. Then there will be one hive here and there where there is a doubtful queen which will need checking. It may be a young queen that we wish to

Sainfoin flower

make sure has come on to the lay, and is laying fertile eggs, or it could be that we will bring one of our home raised queens from the breeding apiary, to replace an old queen who is well passed her prime and needs to be taken out.

The weather is settled and the sun is well up and warming the heaths, just see how active the bees are, it is just the day for us to take you round the hives. I could not have asked for better conditions, there must be a heavy honey flow, as in the Spring the bees hang in the air before the hives, they are heavy with nectar. The damp warmth seems to surge up through the thick straggling stems of these old ling plants, and if you look over there, there is a patch of bell heather still in full flower, the bees did well on the bell a fortnight ago, but now that the ling is flowing they have forsaken it. If you study a comb of honey next month, you will find the patch of bell honey in the centre of the comb, and it will be surrounded by the ling. Down here in Dorset there is the cross leaved heather, a paler version of the common bell, with a different leaf formation, and if you come over here, I will show you a clump of the rare Dorset heath. This is like a huge bell heather, and in dry weather our bees cannot reach the nectar, but nature is kind, for their cousins the bumble bees will bite a hole at the base of the large bell flower and help themselves. This will mean that at a later date the honey bee can take any further nectar via the same hole. As I told you earlier, this co-operation is also seen with field beans in the same dry conditions. Over at this stand there is a stock suffering from paralysis, just look at that pile of dead bees, this is the hive that we have decided to re-queen, so we have got to find the old queen and kill her, and replace her with this fine young queen that I have in the cage. Just a moment, look down there in the heather, there is a lovely male sand lizard, his green slender body blending with the heather shoots, he does not move, it is warm down there and there is a ready supply of food from the dead and dying bees that are suffering from paralysis. He sees us and hurries off into the 'jungle', I have quite often found common lizards sitting on top of crown boards when I have taken the hive roof off to examine a stock, they must find it warm and safe, for with our hives there is no feeder hole to let the bees through. This stock has very little honey in its super, but that is to be expected considering its problems, on the other hand, the hive next door is like a block of lead, the super is solid, the aroma of ling honey hangs in the air, and the bees race forth while the bonanza lasts. There is new snow white brace comb everywhere, and each little wax pocket has a ruby orange drop stored at its base. This is a heather flow at its best, but they rarely last long, vast quantities of honey may flow into the hive for three to five days, but then very often, it is all over and we will have to wait for another year, in fact in cold wet seasons it may never happen at all.

We will take the vehicle up to that little hill top and have our lunch, there is a fine view of Poole Harbour one way, and the heather hillsides the other. We usually stop here for half an hour before we check the last hives off the edge of the firewood, over there you can just see them, a row of little square boxes.

Last week when we were here a Hobby came across those trees and glided right up to where we were sitting, he only saw us at the last moment and swung away over the bank. They breed in this part of Dorset, and if you are very lucky you may even see them chasing the large populations of dragon flies that also live around the fresh water pools scattered about.

After lunch we will drive across to the coast and walk down to Chapmans Pool, just as soon as we have checked this last batch of bees. The coast will be a bit cooler and I think that the afternoon will be very hot with this sun. It will only take twenty minutes to get there and there will be plenty to see.

On these chalk downs of Dorset, beyond the village of Corf, you reach another kind of country; from Poole Harbour, with its sandy soils, we go to pure chalk, and all the softness that goes with it. The walk down to Chapman's Pool is a good mile, and the hillsides are full of flowers and butterflies long since gone from other parts of the land. These Downs have never seen a spray, the meadow flowers are still all there, and only last year, I counted fourteen different butterflies on my walk. There were chalk hill blues, marbled whites, skippers and even a white admiral. I also saw several moths, Dorset is still a haven for moths, and the sight of a Humming Bird Hawk Moth made my day as he dashed from plant to plant. It all goes to show that if the flowers are left alone the rest will follow, but once you take the flowers away, then there is no food for the caterpillars, of both moth and butterfly; and if you spray an insecticide, then all will go, in recent years they have started to plough and sow the tops of the Downs, but by the grace of common sense the sides of the Down are too steep to plough and spoil.

Chapman's Pool to me is as unspoilt a corner of the south coast as I know, and after a day with the bees, it is a grand place to go and just sit on the hill and look out to sea. It is all just as it was a hundred years ago, and except for the fishing huts, it might be unchanged for a thousand. In its way it is rather a forbidding spot, with its blue shale cliffs circling the pool, and the rugged clusters of rocks scattered about the entrance.

Its nearly fifty years since I first clambered down these cliffs with my father, and yet his sketches are almost rock for rock the same. There is something very restful in sitting and watching a scene like this, it is so uncluttered, it gives a sense of permanence that is completely lacking in our hurried existence.

Just look at those Kestrels hanging along the hillside, they do not hover, they have learnt to judge the wind that drifts in from the sea on a hot day. They hold their wings with barely a quiver and watch the Downs below for the movement of a Vole in the tufted grass. At this time of the year I have seen a whole family hunting together along this hill, I once counted seven in a row, right along the Down.

If you have got the energy after all our exertions of this morning, we will climb down to the shore, there is much to see, and I am sure the exercise will do us good. Down there on the beach there are stacks of rotting seaweed, which were thrown up by the rough sea a week ago. Few people realise how much the birds and fish rely on this rotting seaweed. It becomes full of the eggs of the seaweed fly, and the maggots of this fly eat up the seaweed and hatch out into flies, and these flies will lay their eggs in the next lot of seaweed to get thrown back on the beach. On the rocks the rock Pippits wait and search, they live on the seaweed fly, so do Wagtails, not to mention all the Swallows, Martins and Swifts, which stock up on this abundant food just before they leave our shores for the winter. On occasions the sea will come in and wash the rotting seaweed out to sea, millions of maggots will float on the tide and the weed sinks. This leaves a feast ready to be sucked from the surface by the hungry Bass and Grey Mullet. I have seen hundreds of these fish swimming along the shore with their mouths open, filling their crops with food; for Grey Mullet do actually have a crop. Just here from this rock I once took a huge mullet of five pounds or more, and he ran my line out to that clump of weed lying on the surface some fifty yards out.

Today I hear of councils on the south coast wishing to spray our beaches with insecticides to kill these flies, they say the flies spoil the environment for the trippers. How much more will they spoil the environment for the Swallows, Martins and Swifts, who will eat the flies, and may be never reach their destination in Africa. Two years ago I was in Ethiopia watching the wonderful migration of hundreds of thousands of Swallows passing through on their way south. It was a most impressive sight, what a pity to think that for the sake of a few hours on the beach we would make Swallows ill.

Enough of my hobby horses, we will take our shoes off and wander around in the rock pools, it will cool our feet, and there is so much to watch in the world below the sea. The tide is running out, there was a full moon only two days ago, you will remember how bright it was when we moved those bees back from the white clover pollination. That means that the bay here will almost empty, and we can walk through the rock pools, and if we had a net, fill our pockets will prawns. As the tide turns and begins to flow back the prawns will come out from under the rocks and drift in shore with the flow, feeding as they go.

If you follow the coastal path, you will eventually come to the Army ranges, where the countryside is protected by the Army who amuse themselves by firing high explosive at it. In a perversive way this tends to keep people away, and this is good for nature. The flowers of long ago are still there, it is a reservoir of butterflies and moths, and for bees in the right season, there could be a perfect summer flow. In southern England the roe buck has made a fine come back, and its population has never been higher, certainly for five hundred years. They like the peace of the ranges, and right the way back through to Arne we often see them browsing at dawn or dusk. The Seeker Deer is now also fairly common, and only the other day we watched a buck grazing on the salt marsh. When he noticed that he was being watched he sunk ever so slowly into the thick marsh grass, the closer I walked up the deeper he sank, until only the tips of his antlers were visible. I was all for walking right up to him, but my son said "Leave him alone, why upset him".

There being no answer to this statement I stopped about ten yards from him, but it was very interesting to see how close he would sit, and shows how often you may be walking close to deer without ever knowing that they are there.

I think that it is time that we were climbing back up the hill, it will take nearly half an hour to get right to the top and it will make us puff long before we reach the vehicles. As a boy I used to run down the Downs barefoot and back up, at the same pace, leaving my father far behind, but today you will have to wait for me. I will have to find an excuse to stop to watch a buzzard, or stare back out to sea, it is surprising how steep these cliffs have got these last ten years.

I trust you have enjoyed your day out with us, it has been a long morning and an active afternoon, and I know I will sleep tonight, but as I always say "It beats sitting in an office, and I for one spent too much of my young life doing just that".

From the top of the Downs we can look back out to sea, there is a haze hanging over the placid water, it should be another good day tomorrow, and with a bit of luck the flow will continue.

At the end of the summer, when most bee farmers are thinking of the heather, and getting their bees off to the moors, there is another crop to be found on specialised parts of our coast. This crop is called sea lavender, and grows around the edges of estuaries, it gives a fine honey crop, and can produce some very good honey indeed. The problem with it is the question of when it is going to yield, you see unlike most honey crops which are governed by the amount of rainfall we have had, sea lavender flows are governed by the timing of the high spring tides. For me the problems of tides have never bothered my thinking, I have spent too long on too many deserted coastlines not to know what is happening, but to some the tides are a mystery. For this reason I thought that I would tell you a little about this vexing question so that in the future you do not get your feet wet, as I have on many occasions.

Spring tides are not something that happen each spring, but actually happen once a fortnight, it is an interaction of the moon, the sun, and the sea. When the moon is full, or should I say one day after, we come on to the highest tides of the fortnight, and again when the moon is new we also get the highest tides of that fortnight. In between these moon phases the tides get progressively less active, until a week after the high spring tide. Then as they approach the next phase they get higher and higher to another peak; this means that every fortnight you can expect a very high tide, and if the wind is blowing with the tide it many be very high indeed. If the wind is blowing against the tide it will not peak so high.

The problem with the sea lavender is, it grows along the high spring tide mark of the sea shore on mud flats, and it does not like getting a soaking from the high spring tide. If it starts to yield honey during a nice hot spell in August it will be quite happy to go on yielding as long as the next spring tide does not come along and cover it with an hour or two of cold salt water. If this happens then it will stop yielding, and who can blame it. As you can see the secret of getting sea lavender honey is to catch a honey flow between the two high spring tides, if you get there just as the flowers are coming out, and you get a fine settled spell with an off-shore breeze, then it is possible to get a whole month of potential honey yield, but if the weather once turns and douses the flowers, then they may take a week to dry out and get back into a flow situation. That week could be the one when the sun is shining and hot weather prevails. It is like so many of our honey plants, a little obtuse.

Chapter 9

September

In September there is another factor that influences our tides, and that is called the equinoxes, this is the time of year when the earth is parallel with the sun, and the pull of the tides reach their peak. It is often at this time of the year that we get the highest and lowest tides of the year, and it can be a dangerous time for he who does not really know what he is doing.

Going back more years than I care to remember, when I was still in my mid twenties, I went to Scotland on a duck shooting holiday one September. I took with me a Kayak, which I had built the last winter, to make it easier to wander about the Tay Estuary. One glorious September morning I was up at cock crow, and while the long shadows of dawn were still creeping through the reed beds I set off for the great dog bank that runs down the middle of the Tay. It was past low tide, and the river was well on its way running back up the estuary towards Perth. I paddled across the intervening tidal flow and pulled my Kayak well up onto the mid bank, in fact I pulled a good fifty yards clear of the sea, I pushed the paddle into the mud to half the depth of the blade and put a clove hitch over the paddle so that if the tide did come in my boat would not float away.

As I walked down the dog bank a fresh easterly breeze sprang up and blew in my face, I went a good mile and set myself down in the middle of a clump of reeds to await the dawn flight of ducks. I was on my own and the empty estuary stretched out all around me. Soon the teal started to move up the bank and, like little arrows, they came streaking along the mud with the wind under their tails. For forty minutes the excitement was totally absorbing. I must have fired a box of cartridges and I think I picked up two drake teal. The sun was now peeping over the distant hills and sparkling on the blue sea, but the tide was coming in a lot faster than I had bargained for; I looked back the length of the bank and I could see in the distance that my little Kayak was already afloat and bobbing about on the rising sea. It was high time to get back to it, I had forgotten that this was a spring tide and the sea was running with the breeze. I started to hurry but the mud clung to my waders and sucked me back, I began to flounder along and the sweat on my brow trickled down my face washing away the splattered mud. I was soon right out of breath and the boat was still a quarter of a mile away, the sea was all around it and washing further up the bank. I put on a final spurt trying to take a grip of what I can only describe as panic. I battled on assured in the fact that at least the boat was tied to the paddle and the paddle was securely stuck in the mud. Eventually I reached the edge of the sea and could just see the top of the paddle, I pulled up my waders and strod into the sea, and not a moment too soon, the clove hitch had floated up the paddle and was lapping right at the very top of the handle, another wave and it would have been off, and away would have gone my boat. I struggled over the gunnel and fell exhausted into the vessel, I pulled up the paddle and drifted away out across the sea.

Another minute and I would have had to swim across that dangerous stretch of water, and I would have been very lucky to make it. So remember, don't take chances with the tide, or better still don't go shooting teal when it might mean the survival of the fittest.

There is a moment in the tide of bee-keeping when we realize that the summer is over. One can never be sure quite what it is, but it may be a change in the light that tells us that it is now September, and the days are really getting shorter. It can be a sudden gust of wind that cuts through the willows with an edge that we have not noticed before, or the fading of the leaves in the woodland, as the very first tinge of autumn makes you look over your shoulder and think that all is not what it was. A moment later we look back at the hives, the bees are just as busy, the pollen is still coming, but the hives look strangely shorn without their supers. It is time to feed, the cycle of summer is over, the tide is running out far faster than we had realized, and we must get those last two gallons of food into each hive as fast as we can.

I heard a chiff chaff calling in the garden the other day, but his song seems sadly out of place at this time of year. The Swallows are gathering in their throngs, and sweeping back and forth over the gravel pits. The crop is home; and there it stands a row of six hundred weight drums, it never fails to amaze me as I role a drum on to the trailer, to take it to the packers, that the bees have actually picked up and carried this great weight. To think that for every pound of honey that we take to the packers, the bees have picked up about six and we have got ten tons here sitting in the yard. It is surprising how much honey there must be out there in the fields when we think of all the insects that are relying on it for their livelihoods. It must amount to thousands and thousands of tons; I don't think people appreciate the intricacies of the system, for it is not just the bees that need nectar, but hundreds of other insects who take a sip from time to time!

There is a softness about September that belies the real future; we forget about the early frosts of October, and begin to think that they will not happen. "We will feed the bees next week, that will be plenty early enough", but we fool ourselves, for time is not waiting for us, but is hurrying by. The huge population in our hives in August have worked themselves to death, and the sooner we replace the nectar we have taken with sugar syrup the better.

In the dusky evenings of late August and early September, we become aware of whispering wings in the falling light. Families of ducks are on the move; it is a sign of the changing season. During the summer ducks are rearing broods, while drakes loose their flight feathers and hang around the trailing willows, to creep out of sight as we approach.

I have a great affections for ducks, which for a man who has caused the death of several hundred ducks, may seem rather odd. As time goes by the more I watch them and learn about them, the less I wish to raise the gun. If you stand by the flight pond at dusk, shapes will gather in the sky where they will circle and call and, if you reply they will set their wings and drift into the pond. I learnt to call ducks many years ago; first it was the Wigeon on the salt marshes of the wash, then I found that I could move Mallard from their flight lines on the Tay. As the years pass I am happier to watch, to see the movement, to judge the wind and tide, or the moon, and see if I am right. When the moment arrives to lift the gun, when the black silhouettes come over head on set wings, I no longer have the same desire to kill. I know the result, and perhaps it is not a pleasant one, one moment the ducks come into view floating over the willows or reed beds, the next I am picking up a body, his orange paddles still flicking in the half light, and I ask myself what instinct made me pull the trigger.

Thirty years ago I visited the Shetland Islands, I took my gun, and I walked through the heather, it was mid September. It was just as Negley Farson described it in his book, "Going Fishing", a paradox of savage beauty. It was wild and deserted, I hardly saw another man all day, just the odd shepherd on the horizon, the Grouse broke from my feet once or twice, and I put two in my pocket, then I came to a little Loch high in the wind-swept hills. It was late in the evening and I settled down in the reeds to see if any ducks would come; there was a hiss of wings and six Teal were overhead, like little arrows they came out of the sunset, one dropped at the water's edge. Five minutes later the Mallard started to arrive, it is interesting how the Teal always arrive first, they seem to move off to their feeding grounds just before the Mallard. That evening I picked up my Teal and two Mallard and set off in the dark on a long walk home, with the distant call of sea birds ringing in my ears and the chatter of sheep on the hillside. This was the first time I had shot ducks and I have to say that like the drug addict I find it less and less exciting. On that occasion, in a wild land, all on my own, listening to the Curlew and the clamour of the Red Throated Diver, there was a true sense of isolation, of being in a world without man and the excitement was finely pitched. Today such places are rarer and rarer, ducks on the water meadows in the autumn dusk are not the same.

Before we leave the Shetlands and I take you back to some hard work preparing the hives for the winter, I would love you to come and look at the Ravens with me. The Ravens are indeed the arch villains of the wild country of the north. In the Shetlands they are still quite common and can be watched floating across the rolling hills. They are a marvellous sight as they tumble from the sky in display, croaking and beating their way back up into the blue. I once fished a small sea pool for sea Trout, a burn crept in under the cliffs to this pool, and was fed by a cascade of water twenty feet above, the fish came into the pool but could not run any higher. I stood on a rock and cast a fly across the pool, it was March, my feet were cold, and so were my hands, it was a raw day. After about ten minutes, in which time I had caught one three quarter pound sea Trout, I looked up at the cliff above; there sitting on her nest, not thirty feet away, was a hen Raven, her black beady eyes never left me for a moment, but up until then she knew I had not seen her. Once she became aware that I was watching her she became restless, and after a few moments she spread her wings and with a loud croak, swept past the top of my head. I climbed the little cliff and there in the heather was quite a substantial nest with four blue green blotched eggs, that must have been very near hatching for her to sit so tight.

A year or two later I was back in Shetland in September, and I went to North Row, at the very top of the North Island. It is one of the high points of Shetland, with a spectacular view across the Yell Sound and all along the north coast. I stood spellbound by the grandeur of the scene, I looked down at my feet where the cliffs plummeted away to the shore and the Atlantic rollers pounding the rocks. All along the

Buzzard over Dartmoor

cliffs thousands of sea birds swept back and forth, clamouring to each other as they went. On the bright green turf the Whimbrels ran and called, soon they would be away on their migration.

This was the country of the Ravens, and far to the south I saw a band of these robbers flying gently towards me, buffeted by the wind. So I lay on the grass and waited, Ravens can not resist a motionless body on a hillside, I lay as still as I could and watched them turn my way to investigate; soon they were croaking above me, sailing back and forth as they made up their minds whether I was safe to approach. If ever the devil made a bird it must have been the Raven, with the wicked bright eyes, and the huge pick-axe of a beak for taking out eyes of sheep. There was a sighing of wings and a black shadow settled on the rock to my right, he sat there and watched me, enough was enough, I felt very uneasy and sat up and looked at him. He opened his wings, which must have been four feet across, and protested; a minute later he was a black speck beating his way after the rest of the family.

The ploy had worked and proved to me how the keepers of old had ensnared Ravens to come within the reach of their guns. They would 'lie-out' and with the gun at their side let the birds come in, for chasing Ravens on foot would be a fool's game. I returned down the hill to a stone cottage for a cup of tea and left the birds to their hills, perhaps it was safer to watch Ravens from afar.

September is the month of maturing fruit and in our plum orchard I have seen thousands of half rotted plums lying around among the trees. The world knows how welcome such a sight is to wasps, and we get our fair share of them, but what many people don't know is that in certain conditions they are also very popular with our bees. Our orchard is our breeding apiary and we often have two dozen stocks in residence. In September they love to suck at the juice of the plums and will feed off plums still on the trees where the wasps have already nibbled a hole. I am not exactly sure what they do with fruit juice, but I am sure that as a current food for brood, when there is little nectar about, it must be very good. Whether it can be stored over winter without fermenting seems unlikely.

Wasps are much maligned as an insect and do a great deal of good around the garden, I have watched them bring caterpillars back to their nest nearly as big as they are, and I once saw one pick up a wax moth caterpillar from the entrance block of a hive, just as I turned it to shut up the bees; she bit firmly into the middle of the soft white body and with it still kicking at each end, left for home. Having said that even so, as a honey farmer I do not approve of wasps, for they can do a lot of harm to wintering stocks of bees; they will rob colonies when the weather is too cold for the bees to be out, but still quite warm enough for the wasps. They will even try to rob colonies when the bees are flying. In some mild years this robbing may go on right up to Christmas and much of the winter stores will be taken.

If you have an apiary of say a dozen hives, the wasps will try to steal from each hive in turn until they find which hive is least well defended. Then once they have established the weakest, or should we say least aggressive stock, they will rob it right out if they can.

Here is an illustration of the survival of the fittest in action, and all those bee keepers who say "My bees are non-aggressive, I never have any trouble with them", should be warned. Nature has her own way of getting back at us, and really good tempered bees may not be there the next spring for the bee keeper who likes handling his bees without gloves!

When we go round to winter our bees we turn the crown board to give the bees a top ventilation if this is done late on we will find queen wasps hibernating over the crown board, and in most years we kill some two hundred queen wasps on this circuit of the bees. It has the effect of reducing the number of nests near to the apiary, and so likely to be a real problem. Usually we find one or two under any roof, and maybe half a dozen in any single apiary. Three years ago I went to twelve hives on a hill in a wood near Henley, and believe it or not I killed one hundred and twenty six queen wasps on that one visit, and six more the next time I went there. This gives some idea of the numbers of queen wasps that could be over-wintering in the area. On one occasion in recent years down near Hungerford, my son lifted a roof and stepped back in horror, for there on the crown board was a queen hornet; she opened her wings and took off, but for some moments she hovered in front of us as she orientated, like a small flying sausage, and then she flew away through the wood. On the next visit we looked for her, but she never came back; hornets are rare these days, but like wasps if you leave them alone they will leave you alone.

Chapter 10

October

There comes a day in October when the wind is racing through the trees, and tugging the leaves from the branches and tossing them far across the sky, then I know the season is over. It is no longer any good pretending that we are going to see any more days when the bees will be able to get out and collect. Winter is just around the corner and the first frosts will be on the ground before we know it. At this time it is imperative to get round the bee-farm and check that all is fixed for the months to come. By now all the hives will have been fed, though there may be a few feeders to be collected from stocks that were on the heather, and were thus given their last gallon rather late.

Every autumn people will say, "Aren't the leaves staying on the trees a long time this year?", and I will reply that I do not think that it is the case. I watch the trees, I know when the fall will come, and from year to year it is almost always the same. At the end of October there are still thick clusters of foliage on most of the woods, to the extent that you cannot see through the wood, but by the 12th November they have all gone. The trees stand stark and bare, only the oaks still hang on for a few more weeks, I walk through the apiaries and look up at the sky, I watch the leaves drifting like snow over the fields in the late afternoon. I will sit on a hive and see the pigeons come home to roost. It will be many more months before we see the bees again, maybe the odd cleansing flight on a warm still day, but from now onwards they will have to look after themselves, there is little we can do to help them.

I remember just such an afternoon several years ago, when I had gone to Bowsey Hill to take off the last two feeds that had got left on weak stocks that had taken too long to take down their winter food. I put the feeders on the ground and sat on a hive, the apiary is on a little hill in a thick wood, and I looked down amongst the trees. There ambling slowly towards me, his nose to the ground, was an old dog fox; he sniffed a leaf here, put his nose down a hole, walked a few more paces, stopped and looked ahead, but he never saw me. Closer

Barn Own and Corn Rick

and closer he came, just like any old dog out for an afternoon walk. The ground was wet and the scent was poor, he came within twenty feet of me until I was sitting right above him, then he turned and went away through the bushes never knowing that I was there. Only once before in my life had I been near to a wild fox, and that was when I was a very small boy of seven. During the war I ran wild myself, no-one ever knew where I was, and the fields around Warfield hid me for hours. On this occasion I was out one autumn afternoon; it was a lovely sleepy warm Indian summer day, and for interest I climbed up to the bowl of a huge pollard willow. As I pulled my face level with the clusters of branches I became aware of two beady eyes looking back at me, and for a moment we looked straight into each others eyes. There was no sign of fear in the fox, but he looked mildly annoyed at my presence and sprang right over my head, his brush almost touching my face, and moments later as I turned round I could see him cantering off up the hedgerow. There amongst the branches was a warm comfortable lair with a view of all that went on below, and I thought of the many rabbits that might have come all unsuspecting along the hedge and under the willow. What an easy way for him to collect his dinner.

In those days of the early forties the Garth hunt would chase the foxes of that part of the world, and I as a small boy would be taken on my Shetland pony to the Meet. I well remember seeing Colonel Barker waiting on a restless horse as the Meet gathered, and being told that he was 78 years old, and still rode to the hounds while his young daughter controlled the pack. I had a number of elder cousins who would take me on a leading rein and we would try and keep up with the field; this was of course impossible and we would always get left far behind.

The sense of excitement when a wood was drawn still stays with me. The horses stood snorting and puffing, the hounds' primeval notes filled the air, and the crack of whips, the creak of leather, the sucking of hooves being pulled from the soft earth. Then suddenly a fox would break cover and with a thunder of action and noise the whole field would gallop away leaving me and a cousin to try and keep up. There were many foxes in Berkshire in those days, and I never remember us not finding at least one. As a method of controlling foxes, hunting would seem not to be very cost effective, but in the days of open countryside and empty roads, it was a great day out. On cold days we would crash through the frozen puddles in gateways and pound over the hard hollow turf; my cheeks would glow and my toes would become frozen. A hare would streak across the field before us and the Rooks would rise in protest as we passed, wheeling higher and higher into the sky with their brothers the Jackdaws. At the end of the day we would return home with every muscle aching and a memory of a living countryside.

These days foxes are shot or trapped and the hunting habit has died away, we have destroyed that wild land of the forties, and people say we have made it a much better place, but when I think back to those days I wonder.

I am told that in this part of the World there are eight months when the days are dryer rather than wet, and there are four months when the days are wetter rather than dry. I do not mean that it rains more, but that the air remains damp, and so do the interiors of our hives, so from a bee farmers point of view, winter damp will be our worst enemy. This being so the dryer we keep our hives the better. It is not the cold that will kill our bees, but the continuous damp running down the insides of our brood bodies. The battle against this damp becomes the number one priority for the wintering of bees, the use of reversible crown boards with ventilation strips are a must, and in October these will all be turned over before we leave the bees for the next four months.

Chapter 11

November

November is the month when a bee farmer embarks on a week or two, or even three, of hard labour. It is the month of what I call, hedging and ditching. On a bee farm there may be anything up to a couple of dozen out apiaries, and all these have to be maintained. There are over-hanging branches to be cut back, dead trees to be cut out, new fences to be erected, gate posts to be dug in and entrances to sites to be made good. It is best to get all this work done before the real hard weather arrives, and November is a good month to get stuck into it. All our hives are set on stands in pairs, and to each pair of hives there are six supporting pegs, two rails and two end securing batons, this gives a timber square held steady by six oak pegs hammered into the ground. The setting up of new sites, and the relaying of old stands is carried out with zest in November, but if it is left until January, when there is frost in the ground, then the work is very hard, pegs break and split and much bad language is used.

Over the years I have re-laid many hundreds of stands, and there is nothing to beat a fine day out with a sledge, a spike for hole making and plenty of hot coffee. It is certain to get the circulation going, raise the heart rate, and dispel all feelings of tension. Over the last few years my son and I have re-laid every apiary that we have with new oak pegs well creosoted, and set at least a foot or over in the ground. In soft soil or clay the work is easy, but given a site in gravel the work is long and hard with too many broken pegs.

This is the time of year for gloves and two pullovers, but I am sure you will have discarded both by the end of the day. Let us put up a gate post you might say, it will support the fence round these dozen hives. I have put up too many gate posts, Harry Wickens always insisted that we went down at least two feet for a gate post, and set it in with half bricks wedged down with a heavy pole. These gate posts are still standing, and only rot will move them, for they give more than a firm purchase for the fencing wire.

November is also the month of the Pheasant, and nowadays the countryside echoes with the sound of gun fire most days of the week. Harry Wickens and I were once peppered with shot while tending our hives in a quiet corner, by a walking gun who took a shot at a hen Pheasant coming our way. Since the war the number of Pheasants shot has increased steadily, as year by year more and more farms have taken up the sport to create revenue. Nowadays we see hundreds of young birds hurrying about the fields as we drive round tidying up the sites, and I have even seen several hives that have been heavily peppered with shot. I can only imagine that unless our hives fly in the winter, that someone has been taking a low shot at the game.

Many years ago I went to the Tay for a wild fowling holiday, and as you already know, I am very fond of that part of the world. I was then in my early twenties, which makes it a good thirty years back. The object was to enjoy a wild Goose chase, the result was fresh air, miles and miles of walking, hours and hours of sitting in cold damp reed beds, and maybe a Goose or two to take home. One morning as I walked up the sea wall on the North side of the estuary, at the hour of dawn, a grand cock Pheasant rose from the edge of the reed bed and set off across the fields towards a distant copse. It had been a cold night, and no doubt he had come down to the reeds for warmth. I watched him go, and when he was a good seventy yards away I fired a charge of number one Goose shot at him, he fell dead in the middle of the field. I walked over and picked him up, what a lovely, truly, wild Pheasant, with a short stout tail. I walked on another half a mile, and the same thing happened again, another cock broke from the reeds and set off for safety; this time my second barrel reached him just before he was out of range. Both those birds were identical in every way; feather for feather they matched, they could have been twins, and they might have been, but every cock Pheasant in those reed beds was the same. Here the species was perfect, and like the wild Mallard on the foreshore, you could not tell them apart. I am told, on good authority, that they belonged to the Earl of Dundee, and I envy him his Pheasants, and ask his forgiveness if he should ever read this page.

The point that I make is that all those birds were truly wild and all came from the same strain they had probably lived there for several hundred years, and developed to survive in that environment. Today as I drive round the bees I see thousands of Pheasant, but not one is the same as another, they are all hand reared birds, and it would take a century of no stocking and wild breeding to get back to the Pheasants that I saw in Scotland all those years ago.

It makes me wonder if this is the right way to treat the wild species of our countryside. The Earl's wild Pheasants knew their environment, they knew how to fly up a gully, low between the reeds and vanish amongst the stalks, never to be seen again.

For many years we have treated our bees in the same stupid way, with far too much in-breeding, with far too much importing, and not enough careful selection of our own strains. I have seen the British strains of bees reduced to a yellow, sickly, in-active foreign mess. For years I have tried to keep my own strain free from importations, but it was not easy with all our neighbours filling their hives with bees from America, Australia or New Zealand.

There is no doubt in my mind that the species that has been reared in a certain environment, for a good number of years, will be the species that will do best in the environment and not some foreigner that has come along in the interests of cost and convenience.

Chapter 12

December

The end of the year, the shortest day, Christmas, the cycle of the seasons have come full circle, and the bleak month of December comes to its dreary end. For most of the month of December we can be certain of no more than a few snow flurries. In the last week of the year all that can change, and if it does it may stay for a very long time, as it did in the winter of sixty two-three.

For bees, snow is not a real danger as long as it does not last too long. If the grip of winter lasts through to March without a let up, then the bee farmer can be in real trouble. At the end of the year our stocks of bees are still pretty strong, and they have not started to dwindle. We go round the apiaries in the last week of the year and usually all is well, we lift the crown board and look into the top of the brood chamber. The bees should be clustered and still, and it may be that you can see no bees at all, there could be so much food stores that they are still deep down amongst the frames. We trudge around in a thick coat and warm gloves just to make sure all is well. The leaden clouds gather over the hills, snowflakes begin to fall, the ground is hard, the snow will settle, and the toe slots of the roe buck who passed this way last night will soon be filled. The birds are silent, they fluff up their feathers to keep warm, there is little to be cheerful about, that song thrush in the holly tree looks more like a tennis ball with legs and a beak. The flakes fall faster and faster, soon we can't see half way across the field, the plough is turning from grey to white. Suddenly a hare breaks from the grass at our feet; he has left it to the last minute, for he thought we might pass him by and leave him warm and snug in his set. Now he races into the gloom, his brown form fading amongst the snowflakes. Hares have become much scarcer in recent years, their populations have plummeted. Some people say that they have suffered from too much spray and that when they get sprayed with chemicals in the course of modern farming, they lick themselves clean and ingest the poison. Others say that much of their food has been ploughed up in our mono culture farming.

Then there is the question of them being shot. Many years ago much of the Berkshire and Hampshire down land was not ever shot over; or if it was, then very occasionally this meant that if hares were shot there was always a huge supply of new hares to move in from the areas not shot. Today more and more of our land is regularly shot over, and the population cannot cope. After the ravages of myxomatosis, the rabbit populations fell to almost zero and the hares seemed to flourish, but today the rabbits are coming back and the hares are reducing. As a young man I walked the Berkshire downs from Wantage to Shefford. I followed the road up over the hill and down back to the Lambourne Valley. I had lunch at the Swan and walked back in the afternoon. On the way across I counted over one hundred dead rabbits on one mile of the road, this was the mid fifties and the disease was at its worst. Before the outbreak the rabbits lived in their thousands on the down land. A shot at dusk would give the impression that the whole of the edge of the field was moving. At dawn rabbits could be seen hopping in every direction. I remember one December afternoon I watched a fox stalking a group of rabbits sitting near their warren; he crept through the dead grass, as low as his tummy would let him, and every so often he would stop and watch a rabbit that might have detected him, or heard a sound. Closer and closer he came, I stood by the gate not daring to breathe, but he was too intent on the rabbits. Suddenly he was on his feet, flying across the intervening patch of grass, but one rabbit never saw him coming until it was too late, there was a muffled squeal as he caught it around its chest, he lifted his head and trotted off, glanced round, but there was not a rabbit in sight.

Twice in my life I have kept a pet wild rabbit, and all I can say is that they made the best pets that I have ever had. Both lived until they were ten years old, which would appear to be the probable life span for this animal. They were both does and became so tame that they would run around the house just like a cat or dog, or stretch out and sleep on my lap by the fire. Their knowledge of the world about them was really quite acute, they could tell whose footstep it was upon the drive, and a strange footfall would cause them to bolt to the safety of their hutch, whereas if it was a member of the family, they would not move from the run. Lumps of sugar were readily eaten, as were strawberries, blackberries and cherries, but food that was not of the English countryside was ignored. If any food had been sprayed it would be totally rejected.

Tonight the rabbits will come out on this fresh snow; they will trek for hundreds of yards across the fields under cover of darkness,

Rabbits in snow

and it is only after snow that we will be able to see how far they travel. The snow will keep them from the winter wheat, and much of their other food, tonight a broken ash branch will be stripped and the last of the bramble leaves will be picked off the trailing briers.

There was a time when rabbits were very much part of the food of the country people. Every boy on the farm would know how to take a rabbit in a snare, and during the war, when food was scarce I learnt to take them on the high field where they slipped through the fence. There was a boy who tended the cows at the farm and he led me astray, to the distraction of my mother. I remember on a frosty morning we walked over several fields to check his handy work and had collected some six rabbits, when he stopped and put a restraining hand on my shoulder "Look there across that field, can you see something white hanging on the fence?", I strained my eyes, and there was certainly something, "Well", he continued, "That's a hare, you mark my word". We set off across the field, and sure enough there was the hare, it had become caught in the snare and had jumped backwards over the top of the fence. It was hanging in the open a good two feet from the ground for all to see "Just as well we are the first to get here" commented my teacher "Or we well might not have got him".

We have come full circle. The tide of the seasons is fast ebbing away, and in another week or two we will be back where we started. In December the air is cold, fog hangs in the valleys and frost clings on the north facing hillsides all day. For the bee farmer life comes to an abrupt halt. There is no reason to go round the bees, all should by then have been taken care of. The woodpecker nets are on, the top ventilation arranged, the grass round the hives has been trimmed, the overhanging branches cut away, the fencing strengthened and the gates repaired. For the next three months all should be well. You can go home and put your feet up before the blazing log fire. All fulltime bee farmers should have a wood fire in the winter; you will collect so much timber as you tidy up your apiaries, that it would be a pity to waste it. You can sit at home with a clear conscience, you can answer the phone at mid-day and no-one will be surprised. You can even write a book, it will pass the time, and you can kid yourself that it could earn you a crust and someone else might actually read it.

Enough of this self pity. I will take you for a walk, it will do us much more good. Today the wind is bracing, which means it will cut your face like a knife. The ground is hard, the frost has not gone away for several days. A few leaves still cling to the oaks, but the wind will not be denied, for every so often one breaks loose and drifts away across the winter wheat.

Look at the pigeons, see how they rise and dip into the wind. They rise up against the gusts and then let the weight of their body take them forward in a downward sweep, only to repeat the process again and again. They are hungry but the fields are cold, the clover is frosted and they won't eat any today. They are away to the woods to search for acorns and beechnuts. It will be much warmer work in the woods amongst the leaves. They will look up at the ivy and see the berries, but now is far too soon for that feast. The ivy has only just finished flowering, our bees were still working it a month ago, the berries are still hard and unripe. They will have to wait until February or March.

We will walk along the river and across the marsh. The frost is not too deep in the ground yet and the snipe will still be on the marsh. See how they move as we walk through the tufts of rushes. If you look carefully you will see that they all sit in the open, they do not, in fact, hide under the thicker reeds. They will be found on those open patches of water and marsh where they can see their enemies coming. That last bird was not a common snipe, but one of the rarer Jack snipe, his beak is shorter and he is half the size. While the common snipe rise into the sky and leave the area calling as they go, the little Jack snipe will usually only fly about two hundred yards and then drop back into the rushes.

One winter as I stood on the river bank a dog startled a snipe on the marsh. He flew straight towards me. He did not jink as snipe usually do, he was huge and at first I thought that he was a woodcock. As he came closer I could see that the markings were not quite right for

Rooks in flight, right at the end

a woodcock, and his size was far too great to be a common snipe. He flew right over my head and I got a very good look at him. Later, when I got home, I was able to establish that he must have been a great snipe and was certainly one of the rarest birds that I have ever seen in the Thames Valley.

In case you are wondering, there is a reason that we have come this way today; I am going to take you up the canal to that wood a good half-mile from here. There is a sight that is well worth the walk. It is the main roost of ten thousand jackdaws and rooks, and as they come back to sleep amongst the ivy-clad alders and the willows, you will see a picture that will stay with you for a very long time.

We will walk along the wood and wait. The sun has set and the shadows begin to creep across the field, the wind in the trees has dropped a little, but the darkening clouds still drift swiftly over the broken sky. In the distance we can hear the approaching hordes, first it is the high pitched yap of the jackdaws that echoes over the trees, and soon we can hear the steady cawing of the rooks. Here they come, the sky is thick with them, rolling over the trees in a continuous wave, all calling to each other as they drop into the tree tops. It is a grand sight and it lasts for about ten minutes, eventually the whole huge family is home for the night. They sit in the trees still chattering away to each other. Slowly the noise begins to subside and they settle down to sleep. The shadows close in, silence descents on the wood, until you imagine that there is not a bird in the world.

It is time to go. The last streaks of daylight hang on the horizon, a pair of ducks passes high overhead, and a tawny owl breaks the silence from the woods on the hill. There is a whisper of wings above our heads and the plaintive cry of a plover pierces the dark. The last bird to go to roost has passed, it is time we went home to that warming fire and left the fields and woods to the people of the night.

Conclusions

"Every book must have a conclusion", the words of my father many years ago, so mine will be no exception to his wish! "Is there a PURPOSE?" I pose the question asked by that eloquent writer William Caine at the turn of the century. "Is there a purpose to life?" I ask the question again. I believe there is, I believe that we are all part of an intricate, tightly woven pattern of life, all life relies on all life, each part leaning on the other. I hope that I have taken you to some lovely parts of the country, and shown you what I have seen and enjoyed, and lived for. I have made my living from the fields, the woods, the river banks, and the hedgerows, all of which I feel part of. I may have taken from the countryside six to a dozen tons of honey each season, which was a surplus that was available, I hope that I have not damaged the fabric as I passed.

If having read my book, you have come to the opinion that I am a natural predator, then you are probably right and I must plead guilty. On the other hand I would like to feel that I have lived within the bounds of the natural world that supports me. The problem today is, we are no longer living within the same natural world that we did fifty years ago. We are now damaging, changing, and destroying that world, with ever greater haste.

Our politicians tell us that we must have an ever expanding economy, that we must create more and more wealth. Our cities absorb a greater and greater proportion of nature's output, and in return pour back vast quantities of poison, this just cannot go on, there must be a finite limit, and many feel that it is fast approaching. The very threads of the PURPOSE of life are being destroyed, talk about the Midas touch, I believe that the fairy tale is about to become true. If all we touch is turned to gold, what will we have left, maybe vast capital assets, but at what cost, dead rivers, dead trees, dead birds, and for me, no wild flowers and hence no bees!

"Something has to be done", I hear you say, "Someone has got to put on the brake", the trouble is, none of us are wearing seat belts. Politicians dare not touch the brake for fear of the outcry of the injured section of the community, who do not want to give up their profligate

Bumblebee on Dandelion

ways. Already some of our cities are grinding to a halt, and why, because nature cannot afford their up-keep, and neither can the people who live in them.

We all look back to our childhood through rose tinted spectacles, but I firmly believe that when I was a boy, the countryside was in balance. Each little piece supporting all the other little pieces, today there are great holes in the cloth, and as a bee farmer I can see the steady destruction day-by-day, there is no let up.

For myself I do not believe in politicians, and this is not the cynicism of old age, I can safely say that I have never put one into office, they have always been going in the wrong direction whenever I have met them. They do not seem to understand my world or the real

PURPOSE of true balance. They are like small children who take their toys to pieces, and then have a tantrum when they cannot put them together again.

There has to be a new thinking, if the purpose is to survive. Does your politician know the difference between a blue winged olive, or a mosquito, can he tell a grasshopper warbler from a cricket, or even a honey bee from a wasp, I doubt it. They do not live in my world, and I certainly do not wish to live in theirs.

The PURPOSE has been around for five thousand million years, it has grown and developed, it has evolved by taking the best from each generation, the strongest, the fittest, the wisest. Now our caring society is to meet head on with Darwin, no longer must we serve the PURPOSE, the PURPOSE must serve us, we can afford to put the rules of evolution behind us. We have conquered nature, we can do without the natural world of the past, we can now order our own flowers, our own animals, our own types of fishes, but I ask, "Is that what we really want?" Or could it be that in our moment of triumph, when the battle against the very PURPOSE of life has been won, we will destroy ourselves. My last question to my readers must be, "What do you want, why not make up your mind before it is too late?"

For myself, you know what I want, you have seen my life style, a natural predator, living in the fields, woods and hedgerows, by streams and estuaries. If I can take a brace of wild brown trout from a clear clean brook for breakfast, spend the day working a hundred hives on a honey flow, and spend the last hour of light sitting in the shade of some fruit tree, and wait for night to come, then I will not complain.

Finally, what of the PURPOSE? do we need it? I am sure we do. We say there is a PURPOSE behind everything, and it is that PURPOSE that keeps us going forward. Now suppose you should take the PURPOSE away, what then, I ask you, will you still keep going?